The Complete Guide to Using Google in Libraries

The Complete Guide to Using Google in Libraries

Research, User Applications, and Networking

Volume 2

Edited by
Carol Smallwood

ROWMAN & LITTLEFIELD
Lanham • Boulder • New York • London

Published by Rowman & Littlefield
A wholly owned subsidiary of The Rowman & Littlefield Publishing Group, Inc.
4501 Forbes Boulevard, Suite 200, Lanham, Maryland 20706
www.rowman.com

Unit A, Whitacre Mews, 26-34 Stannary Street, London SE11 4AB

British Library Cataloguing in Publication Information Available

Library of Congress Cataloging-in-Publication Data

The complete guide to using Google in libraries / edited by Carol Smallwood.
volumes cm
Includes bibliographical references and index.
Contents: volume 1. Instruction, administration, and staff productivity – volume 2. Research, user applications, and networking.
ISBN 978-1-4422-4689-8 (v. 1 : hardcover) – ISBN 978-1-4422-4690-4 (v. 1 : paperback) – ISBN 978-1-4422-4691-1 (v. 1 : ebook) – ISBN 978-1-4422-4786-4 (v. 2 : hardcover) – ISBN 978-1-4422-4787-1 (v. 2 : paperback) – ISBN 978-1-4422-4788-8 (v. 2 : ebook) 1. Web applications in libraries. 2. Google. I. Smallwood, Carol, 1939– editor.
Z674.75.W67G67 2015
025.042–dc23
2014040494

∞™ The paper used in this publication meets the minimum requirements of American National Standard for Information Sciences Permanence of Paper for Printed Library Materials, ANSI/NISO Z39.48-1992.

Printed in the United States of America

Contents

Foreword

Michael Lesk

I first saw what would become Google in 1996 while visiting Stanford, where the inventors were then graduate students. That PageRank predecessor was called Backrub, and it was only recently being tried on web searches. Until the invention of the PageRank algorithm, the web was often overwhelming—too much information, of indifferent or even appalling quality. Once Sergey Brin and Larry Page developed a way of selecting good web pages, and made it available through their new company, Google, search became useful. Combined with the web, modern search systems are now the dominant way that students find materials for class assignments.

Google has now developed services far beyond text search. Google software will translate languages and support collaborative writing. The chapters in this book look at many Google services, from music to finance, and describe how they can be used by students and other library users. Although this book concentrates on Google offerings, other companies have expanded into online search as well. There are direct competitors such as Microsoft, with its Bing search engine, and with Microsoft Academic Search as an alternative to Google Scholar. In addition, companies in the social networking realm, such as Facebook and Twitter, offer other ways of connecting with people and information. However, the focus of this book is on Google and its products, as they enable new activities by the users of libraries.

Many new kinds of information are now available to patrons. Scholarly papers were mentioned above, and many libraries also subscribe to publisher search engines such as ScienceDirect or SpringerLink. In medicine the PubMed Central service of the National Library of Medicine offers research papers, and most recent ones are available free in full text. Patents are easily searched with either Google or other patent search systems. While journal articles and patents are of interest to serious researchers, Google also supports searching for shopping and health information, and is a leading source of maps and driving directions.

Additional dramatic changes from traditional text are image search and video search. Photograph storage and sharing sites such as Flickr and Google's Picasa provide billions of images, and in addition the search engines index the images they find on web pages. YouTube, another Google offering, has a dominant position in online amateur video, including my own three minutes on how to groom our cat. As with scholarly publishing there is also a large commercial market, such as Netflix and Hulu, which as of this year don't usually offer attractive terms to libraries.

Going beyond information resources, there are now successful collaboration services available from Google and others. You can make conference calls with video and shared screens using Google Hangouts, Skype, Citrix GoToMeeting or Adobe Connect. We have always thought of library research as a solo activity, and in "As We May Think," the original 1945 article that forecast information retrieval and hypertext, Vannevar Bush talked about the new experience of scientists working in teams. Although large-team research is now common, and the papers announcing the Higgs boson had some 3,000 authors each, traditional humanities scholarship remained individual. Writing documents with small numbers of colleagues often involved delays while each author in sequence took over the writing and made edits. Today Google Docs enables multiple people to edit the same document at once. An ingenious use of color lets each participant watch in real time as the other participants edit, and keeps track of who is doing what. If the goal is to create a website rather than to write a report, Google Sites is now one of the most popular platforms. Google is also involved in social networking, with services such as Google+ and Orkut (the latter now discontinued in the United States).

Other tools view social developments over time and space. The Google Trends service, for example, will show you when and where people are searching for topics. Not surprisingly, searches for "swimwear" peak in June and searches for "snowmobile" peak in January. Searches for "kangaroo" are most popular in Australia and searches for "biryani" are most popular in India. The Google "Ngram" addition to book search tells you the numerical history of word usage over time. For example, the word "gout" was most common in eighteenth-century books, "catarrh" in the late nineteenth century, and "tuberculosis" in the early twentieth century. You can see how fleeting fame can be from an Ngram plot for "Anthony Hope" (the author of *The Prisoner of Zenda*).

A disappointment was Google Flu Trends. In 2009 a paper from Google suggested that watching search terms could detect influenza outbreaks before the official statistics. Later observations have not confirmed this and suggest that it may have been a fluke; if you measure twenty different correlations, one of them will be significant at the 0.05 level. Marketing researchers, however, still exploit search data in many contexts, and political campaigns are accused of trying to manipulate Google results so that searches for their opponents will retrieve derogatory information.

This book covers a wide variety of Google services of interest to library patrons. Most people think they are much better at searching than they really are (read about the Dunning-Kruger effect), and almost anyone is likely to learn something useful about Google technologies from these chapters. Whether your problem is managing information, managing people, managing your finances, or just finding entertainment, there is almost certainly some chapter in this book that will suggest how you could be more effective at it.

Preface

Carol Smallwood

Google has vast resources that are often insufficiently applied by library users. As a remedy, this anthology comprising chapters by practicing public, academic, school, and special librarians and LIS faculty in the United States and Canada presents a wealth of creative and practical approaches to utilizing Google applications. The topic of "Google as a library partner or competitor" was suggested to the authors to encourage wide coverage of the various issues. There was no contact with Google in connection with this anthology.

The Complete Guide to Using Google in Libraries, Volume 2: Research, User Applications, and Networking has thirty chapters. The contributors could submit two chapters that totaled three to four thousand words, or one chapter of that length; the previously unpublished chapters could be by one, two, or three contributors. These concise, how-to chapters are based on experience, research, and case studies and bring together an abundance of practical information to aid librarians, library consultants, LIS faculty/students, and technology professionals in their work.

It was a pleasure working with these experienced contributors willing to share with other librarians how Google can meet diverse needs of library users. Their willingness to help colleagues is crucial in our rapidly changing world of technology. A special thanks to Mary Lou Andrews, for technical help.

Acknowledgments

Rose Petralia, instruction librarian, Evans Library at Florida Institute of Technology

Judy Donovan, past president, School Library Division of the Association for Educational Communications and Technology

Judith Wines, director, Altamont Free Library in Altamont, New York

April Ritchie, adult services coordinator, Kenton County Public Library, Erlanger Branch, Erlanger, Kentucky

Melissa Cornwell, distance learning librarian, Norwich University, Northfield, Vermont

David Robinson, adult services librarian, Shenandoah County Library, Edinburg, Virginia

Samantha Helmick, librarian, Burlington Public Library, Burlington, Iowa

Part I

Research

Chapter One

Beyond "Good" and "Bad"

Google as a Crucial Component of Information Literacy

Andrew Walsh

The topic of Google is a common one in academic library instruction sessions, but unfortunately it is often for all the wrong reasons. In today's increasingly complex information landscape, library resources face strong competition from Google due to the latter's easy-to-use nature and ever-improving search results. Google's search algorithms, the computer programs that determine which web pages show up first for a given search, now take hundreds of factors into account when ranking results (Google 2014), and can quickly return massive amounts of information for any conceivable topic. One possible approach to library instruction, then, is to demonstrate why Google and other open-web resources are not as high quality as library subscription databases and other paid-for content. This strategy typically involves teaching that library content is more credible and fact based, that it was written by subject matter experts, and that it has advanced organization to facilitate the discovery of relevant results for a given information need.

But if the overall goal is to get students to bypass search engines and start their searching at the "better" library, this strategy is failing, as an overwhelming percentage of students start with Google despite attending in-person library orientations and completing online library tutorials. Head (2013) conducted research on nearly two thousand college students and found that "Google was students' go-to preferred research resource, whether they were college freshmen (88%), high school students (89%), or college sophomores, juniors, and seniors (87%)." The study also referred to a "deeply ingrained habit of using Google searches" and even that "most were intimidated by the plethora of print and online sources their college libraries offered and uncertain how to access or use them."

For some, the answer to these findings might be increased marketing of library databases. But instead of viewing Google as a competitor, a better approach is to connect library instruction to students' existing research processes, which as we know starts with Google most frequently. By adopting this strategy, Google becomes a natural starting point and a crucial component to effective library instruction. This chapter explores why a Google-centric approach makes more sense than focusing solely on library search systems and provides specific examples that can be used in a classroom setting. The target environment is the undergraduate college level, but the concepts apply to any type of instruction at academic, public, and school libraries and can be adapted as seen fit.

GOOGLE *IS* USEFUL FOR ACADEMIC RESEARCH

Library instruction programs that discourage Google often teach that the search engine can be helpful for personal information needs, such as finding local businesses or reading entertainment news, but for those who need to do serious, academic research, it falls hopelessly short and a new, separate set of skills is required. This strategy is both misleading to students as well as at odds with actual behavior and experience, which results in a significant disconnect. In addition to Head's study (2013) showing overwhelming student use of Google, Anderson (2012) points out that many times even we librarians don't go straight to subscription databases: "The reality is we (librarians) use Google for our own information needs, so why should we expect the students to only use what we want them to use?"

It is also becoming increasingly misleading to suggest that academic information sits only behind database paywalls. Many large repositories and directories for scholarly content are available through a regular Google search including the Directory of Open Access Journals (DOAJ), OAIster, and numerous discipline or institution-specific repositories. Many of these initiatives are investing heavily in Search Engine Optimization (SEO) to specifically improve their items' positioning in Google searches. In addition, Google Scholar provides free access to a wide variety of full-text scholarly content. These resources will likely only grow in usage and importance as the open access movement continues to develop. With the economics of scholarly publication breaking and some academics and librarians even boycotting major publishers and database vendors (Neylon 2014) the shift toward this Google-friendly open access content will only accelerate. An instruction strategy that ignores or discourages Google steers students away from these valuable resources and also fails to educate them on important economic and social issues relating to information production and dissemination.

Many librarians make it their mission to undo the deeply ingrained "Google habit" of students. But if we reject the idea of Google and the library in competition, these same behaviors can be viewed as building blocks: real research practices that, although simplified and incomplete, can be built upon. Purdy and Walker (2013) argue that "academic research practices need to be *connected* to students' existing practices rather than set up as wholly separate from (and better than) them." They refer specifically to English composition classes as spaces in which students, whom they argue have existing research identities and skills, "cross a threshold" into a new academic environment. This idea can be applied to all library instruction, and by not rejecting previous practices involving Google or terming them "unskilled" or "illiterate," students will recognize connections that exist across multiple domains and develop more broadly applicable information competencies.

USING GOOGLE TO GO BEYOND "GOOD" AND "BAD" SOURCES

Librarians who are hesitant to promote the use of Google for academic purposes certainly have good reason: there is plenty of information on the open web that is inaccurate, misleading, or biased. But rather than that being a negative, it is a perfect opportunity to teach critical information literacy skills that apply in multiple research environments. Most librarians are in agreement that the goal of library instruction should be to teach fundamental skills for finding, evaluating, and synthesizing information in a variety of formats, not simply showing which buttons to click in a particular type of search interface such as library databases. According to the Association of College and Research Libraries (ACRL), the division of the American Library Association involved with academic library instruction, information literacy "combines a repertoire of abilities, practices, and dispositions focused on expanding one's understanding of the information ecosystem, with the proficiencies of finding, using and analyzing information, scholarship, and data to answer questions, develop new ones, and create new knowledge" (ACRL Information Literacy Competency Standards for Higher Education Task Force 2014). This definition clearly goes far beyond paid subscription resources and suggests that library instruction must foster more fundamental competencies. And Google, then, becomes an ideal venue to practice the skills found in the ACRL definition for the very reason that it provides a wide variety of information resources of varying credibility levels.

In order to accomplish these broader goals, effective library instruction must go beyond simplistic categorizations of information sources such as "good" vs. "bad" or "credible" vs. "untrustworthy." In the real world, very few sources are completely perfect, and also very few have no value at all (and the ones that do usually make their inadequacies clear even to the "untrained" searcher). When students are younger, such as in middle or high school, it can make sense to oversimplify and not allow the use of Google. Many college professors, however, continue the "blacklisting" strategy and forbid the use of Google for class assignments. This is unwise for several reasons. As previously mentioned, it is unrealistic and disconnected from actual practices (of both students and instructors) and it ignores a growing landscape of high-quality open-access resources. Explicitly forbidding specific resources also completely removes the critical thinking component that is supposedly a central outcome for higher education.

One method of library instruction that does involve Google and is popular among many librarians is the "hoax" lesson. In this activity students are presented with a website that looks legitimate but in reality is much different than it appears. For example, one website frequently used in library instruction over the years (http://zapatopi.net/treeoctopus) presents extensive information about the "Pacific Northwest Tree Octopus," a completely fictitious creature, and urges its protection. Another (www.dhmo.org/facts.html) warns the reader of the dangers of the chemical "dihydrogen monoxide" which additional research reveals to be water. Although this approach typically does elicit a response in students (as well as often fool many of them!) a good lesson on evaluating information should go far beyond the "hoax." The takeaway lesson here is often something to the effect of "see, sites on Google can trick you, so you'd better be safe and just use library databases." But in the real world, students will rarely if ever encounter a blatantly false website or one set up solely to deceive them. What they will encounter, on the other hand, are sources that fall into gray areas, such as websites with some credible information but a potential conflict of interest on the part of the author or supporting organization. Or websites that make some important points but might be out of date or not at an appropriate level for the assignment. Also implicit in the "hoax" lesson takeaway is the idea that library databases are high quality so they do not require evaluation, which is far from the truth.

Because of these realities, librarians must go beyond lessons that place information sources into simplistic binary categories or have the ultimate goal of scaring students away from Google. Instead, instructors must recognize that nearly everyone will and should use Google, at least at the earlier stages of the research process, and that determining which information sources to use and how to use them is a carefully considered decision depending on many factors. The following section includes several specific lessons that feature Google, which is a natural and effective starting point for a variety of information literacy themes ranging from critically evaluating sources to choosing a topic. In addition, instruction on Google search tips and strategies as well as specialized Google search engines helps students become more sophisticated seekers of information.

Throughout these lessons, however, it is crucial to emphasize that although Google is an excellent place to get started, it should never be the only search system in a researcher's repertoire. Instead, students must understand that specific situations call for other resources, which may range from subscription library databases to books in the stacks. The following lessons present several instances where Google and library databases form a formidable "one-two punch," with the former typically used first and the latter afterward to fill in the gaps that remain. In addition to the situations to be described, there are other areas where library databases hold a distinct advantage, including their support for controlled vocabulary, the length and searchability of archives, more precise facets and filtering options, and other features that appeal most strongly to upper-division and graduate-level courses with more substantive research requirements. Despite these advantages of library databases, however, an approach that starts with Google and builds from there is ideal because it helps remove the intimidation and uncertainty toward library resources frequently cited by undergraduates and allows them to more effectively appreciate the distinct and complementary roles that library databases and search engines play.

HOW TO USE GOOGLE IN LIBRARY INSTRUCTION: LESSON IDEAS

The "CRAAP Test" Framework for Evaluating Information

Given that information sources cannot be categorized simply as "good" or "bad," it becomes necessary to ask a variety of questions that dig deeper into whether a particular source is appropriate for a given need. A particularly valuable lesson, then, is to give students a common framework they can use when evaluating any information source, whether found through Google or a library database. A popular such framework is the CRAAP test, designed at CSU-Chico, so named because it identifies Currency, Relevance, Accuracy, Authority, and Purpose (CRAAP) as the five primary evaluation criteria. For each, students are given a set of questions to ask when evaluating their sources.

Currency, for example, asks when a source was created or published and whether it has been updated, while also urging students to think about their own research topic and how recent a source should be to be considered appropriate. For something fast-changing such as social media technologies or an emerging medical treatment procedure, a source two or three years old might not pass the Currency test, while in some arts and humanities fields a source from decades ago might still be the seminal work on the topic. The four other criteria have their own sets of questions, and as a whole the CRAAP test avoids quick and easy answers and is highly context dependent, which helps students develop the broader information literacy competencies mentioned by ACRL. See the entire CRAAP test handout with all of the questions at www.csuchico.edu/lins/handouts/eval_websites.pdf.

It is important to emphasize that the CRAAP test is a framework for evaluating all types of information, not just those found on Google. Library tutorials are sometimes framed as exercises solely to teach "evaluating websites," as if resources in library databases do not have to be carefully examined. These resources do still need to be evaluated, of course, and in library databases the criteria of Currency and Relevance play particularly important roles.

Possible Classroom Activity: After teaching the CRAAP test, put students into groups of four or five and give each group a topic written on a scrap of paper. Have the students search Google for their assigned topic and identify one website they feel would be appropriate for a college research paper and one that they feel would not. Ask them which criteria they used to evaluate the websites and which specific questions they asked.

Topic Generation and Refinement with Google

A common undergraduate assignment is an argumentative essay on an issue of the student's choice. And in order to craft a high-quality paper, it is important to select an appropriate topic. Students face a wide variety of methods for selecting a topic, including consulting books, news sources, or even family and friends. Librarians often attempt to steer students into research databases right from the start, but a more realistic and effective approach is to start where most students already are: Google. A quick Google search allows students to learn background information, browse recent news stories, refine a broad topic into a narrower area of interest, and identify important terminology that can be used as keywords in future searches or provide new directions of focus. Some search results may be less than ideal as a cited reference in the final research product (and using the CRAAP test, especially the Authority criterion, can help determine this), but at this stage they may be perfectly fine for browsing, clarifying ideas, and learning more. This is one example where Google and library databases naturally go together, as ideas and search terms generated from Google searches can be used later in a database.

Possible Classroom Activity: Have students browse for a topic of interest and answer five questions designed to narrow a broad idea into a workable topic: Who is involved in this issue and who will you focus on? What are the important issues at stake? Where is this topic relevant and where are you focusing? When did this topic start and how recent should your sources be? Why have you chosen this topic? Also have students write down keywords as well as synonyms and alternatives for those keywords as they encounter them.

Advanced Searching with Google Search Operators

Google supports much more than just simple keyword searches, and students are often not aware of additional techniques that can yield much more helpful search results in the context of academic research.

- Site operator search (e.g., *Minimum wage site:nytimes.com*): A site operator search limits results to a particular domain name, in the above example the *New York Times* website. The same search can also be done for just a particular domain name extension such as .edu or .gov, which can be an effective way to identify sources that will be more likely to be appropriate for a college paper (and that the CRAAP test can help confirm).
- Filetype operator search (e.g., death penalty filetype:ppt): A filetype operator search returns only results in a particular format, such as PDFs, Word documents, or PowerPoint presentations.
- Excluding terms and (e.g., *"jaguars" -car -nfl*): Keywords can be eliminated from a Google search by entering a hyphen (-) before them. To avoid having Google automatically include synonyms for a search term, put it in quotations, the same method would be used to search for an exact phrase. This example would search for pages containing the exact keyword *jaguar* and omit any results that contain the keyword *car* or *nfl*, so it should return mostly results about the animal, and eliminate the vehicle and football team. The operator can also be used to eliminate results from a particular domain, such as *-site:wikipedia.org*.

Possible Classroom Activity: After explaining search operators and giving students a handout, assign a worksheet asking them to construct various advanced searches utilizing Google search operators. For example, one question in the assignment could be "Search for sources in PDF format about affordable housing from government organizations."

"Vertical" Google Search Engines Including Scholar

In addition to regular web search, searching Google's specialized (also known as "vertical") search tools directly can be useful in the context of a research assignment. Google News, as previously mentioned, can be used to find a research topic that is currently in the public consciousness. YouTube can be used to find documentaries, independent films, and other video material that can be beneficial as sources, and the CRAAP test allows students to critically evaluate the credibility and appropriateness of each one they encounter. Other Google vertical search engines including Books, Images, Blogs, and even Patents may also be useful depending on the source types needed.

Scholar is Google's vertical search engine most effectively geared toward academic research, as it combs the web to find content deemed "scholarly" by an advanced algorithm. Results are varied, ranging from open-access publications to PDF copies on institutional websites, and the main advantage is Google's familiar search interface and intuitive results display.

This is one of the clearest examples in which Google is best used in conjunction with library databases. Some of the resources included in Scholar have only abstracts available for free, and it is common that students will find articles that prompt them to pay for the full text. Often those same articles are available through their library's subscription databases, and once they have identified the material all that remains is a relatively basic known-item search in a discovery service, citation finder, or other service made available for searching across multiple databases. And for some libraries, an institutional affiliation can be specified in Scholar's Settings, which will add a full-text link from the library directly next to Scholar search results.

GOOGLE FOR LIFELONG INFORMATION LITERACY

Another major advantage of an instruction strategy that starts with Google is that the shift in focus from individual search systems to broader information literacy skills ensures that the lessons will be applicable for students for the rest of their lives, not just in their immediate educational surroundings. This is particularly important due to the fact that database access is typically shut off when a student graduates. Most librarians agree that fostering broader information literacy skills is the true mission of academic librarians, not purchasing and subscribing to particular resources. According to Fister (2012), it is "entirely the wrong argument" to say libraries derive their primary value from providing access to articles "that are so much better than what you find online for free." She suggests that we stop "bragging about our shiny tollgates" in favor of open access and broader skills. As information professionals we can start to do this by (1) utilizing an instructional strategy that begins where students already are, Google; (2) recognizing habits they already possess; and (3) gradually adding on new layers. This approach will help students make stronger connections between personal and academic search environments, form a more complete understanding of the information landscape, and ultimately have increased success when dealing with information of all types.

REFERENCES

ACRL Information Literacy Competency Standards for Higher Education Task Force. 2014. "Draft Framework for Information Literacy for Higher Education." *Association of College and Research Libraries.* http://acrl.ala.org/ilstandards/wp-content/uploads/2014/02/Framework-for-IL-for-HE-Draft-1-Part-1.pdf .

Anderson, Mary Alice. 2012. "Google Literacy Lesson Plans: Way Beyond 'Just Google It.'" *Internet@Schools* 19(4): 20–22.

Fister, Barbara. 2012. "Do Librarians Work Hard Enough?" *Inside Higher Ed*, April 2. www.insidehighered.com/blogs/library-babel-fish/do-librarians-work-hard-enough .

Google. 2014."Algorithms." *Inside Search.* Accessed May 23, 2014. www.google.com/intl/en/insidesearch/howsearchworks/algorithms.html .

Head, Alison J. 2013. "Learning the Ropes: How Freshmen Conduct Course Research Once They Enter College." *Project Information Literacy Research Report*, December 4. http://projectinfolit.org/images/pdfs/pil_2013_freshmenstudy_fullreport.pdf .

Neylon, Tyler. 2014. "The Cost of Knowledge." *The Cost of Knowledge.* http://thecostofknowledge.com/ .

Purdy, James P., and Joyce R. Walker. 2013. "Liminal Spaces and Research Identity: The Construction of Introductory Composition Students as Researchers." *Pedagogy: Critical Approaches to Teaching Literature, Language, Composition, and Culture* 13(1): 9–41.

Chapter Two

Enhancing Music Collections

YouTube as an Outreach Tool to Share Historic Sheet Music

Steven Pryor, Therese Zoski Dickman, and Mary Z. Rose

The users, or performers, are out there, and we have to connect to them.

—MacAyeal 2014, 106

With the advent of social media, those with Internet access have turned increasingly to YouTube to watch recorded music performances. As part of a music digitization project to bring historic nineteenth-century sheet music to a larger online audience, the Gordon Colket Sheet Music Collection was digitally scanned. When completed, it became clear that the use of more familiar online tools could increase access to the collection. Selected sheet music titles from the collection were therefore performed on piano and recorded. The edited final recordings were then uploaded to YouTube with a link provided to the digital collection.

BACKGROUND TO THE PROJECT

The Gordon Wright Colket Collection was acquired by the Friends of Lovejoy Library for Southern Illinois University Edwardsville in 1971. Colket, a dealer in illustrative materials, assembled the 120 pieces of sheet music from other dealers, at auction, estate sales, and through trading, primarily in New York City and New Jersey. Most of the pieces are by American composers and date from the nineteenth century. The collection is rich in Civil War vintage pieces.

In an effort to make the Colket Collection more accessible, a digital collection was created in 2013 (SIUE 2014). Twenty-four pieces were omitted from the digital collection because they are either incomplete or under copyright restriction. The remaining ninety-six complete pieces are in the public domain and date from 1832 to 1917. Most have beautifully illustrated covers. Some pieces include dance instructions. The collection includes works by notable composers of the nineteenth century, including Henry C. Work, Francis Brown, George Maywood, and J. H. McNaughton. Several important lithography firms are also represented in the collection. The images in the digital collection were retouched using Adobe Photoshop to increase contrast and brightness for improved usability for performance. The digital collection was created using CONTENTdm software and is a CARLI digital collection, hosted by the College and Academic and Research Libraries in Illinois (CARLI) consortium.

ENHANCING THE DIGITAL COLLECTION BY PERFORMING SELECTED PIECES

As the digital collection was being created, selected piano pieces were performed and recorded by music and fine arts librarian Therese Zoski Dickman. A Zoom H2n Handy digital audio recorder was used. The reasonably priced device (approximately $150) had been highly recommended by music librarian colleagues and did an excellent job. The audio clips were stored as 24-bit/96 kHz WAV files on an SD card within the recording device. The hand-held recorder was mounted on a microphone stand and positioned near the curve of the Yamaha grand piano with raised top for optimal sound capture in the library auditorium. A remote control device enabled the pianist to easily stop and start the recorder when needed. Multiple recordings were taken of sections of a given piece of sheet music. Each "take" was briefly listed by the pianist at the time of performance; an iPad, tablet, or paper could be used for this. Good takes were noted for later consideration when creating an edited final audio file for a given piece. A metronome was consulted when needed to keep the tempo consistent for the various recorded segments. Depending on length, one or two pieces were typically recorded during a two-hour recording session.

The audio clips of a given sheet music piece were downloaded to an iMac computer and transferred to a high-quality audio compact disc. The pianist reviewed the various audio clips and selected the best ones to use for the final audio file of the piece. The exact timing of the chosen clips was noted on a printed copy of the sheet music. Senior library specialist Doug Meyer, a jazz and classical bassist with music recording experience, used the annotated sheet music to splice the identified audio clips together into a final edited audio file. He used the music software editing program Sound Studio. Similar programs, such as Audacity, can be used for PC users.

In July 2013, the audio files for ten works had been prepared. Those were added to the Colket Illustrated Sheet Music digital collection by Mary Rose, catalog and metadata librarian. While usability of the materials had been enhanced by adding the audio clips, access to the audio clips still seemed limited. Given the ready accessibility of YouTube and its popularity, it was decided to harness YouTube to facilitate greater access to the audio clips and the Colket digital collection. A link back to the digital collection would be provided to encourage further use. Steven Pryor, director of digital initiatives and technologies, converted the edited performance audio files to video files. On September 30, 2013, Pryor uploaded the video recordings to YouTube.

LEVERAGING YOUTUBE

Step 1: Reviewing the YouTube Terms of Service

The YouTube Terms of Service were reviewed to ensure that the library and university were not relinquishing any unnecessary ownership rights to these important materials. Indeed, the Terms of Service web page states in part 6C, "For clarity, you retain all of your ownership rights in your Content." YouTube is granted several rights related to duplicating, distributing, and performing the "Content" in any location and any format "in connection with the Service and YouTube's (and its successors' and affiliates') business." Further, it is specified that these rights terminate within a "commercially reasonable" time after deletion if the owner decides to remove the videos ("Terms of Service" 2010). It was agreed that these protections were adequate and appropriate for the sharing of this content.

Step 2: Transforming the Audio Files into Video Files

YouTube does not accept audio files. To upload music to YouTube, each audio file needed to be converted to a video format. This was achieved by adding an image of the cover or title sheet of the digitized sheet music. A simple way to accomplish this was with Windows Live Movie Maker, a free download for Windows computers from the Microsoft website.

Converting an audio track and accompanying image to a video file with Windows Live Movie Maker consists of the following basic steps:

- In a new project, click Add Video and Photos, and choose the desired image file.
- The Add Music button becomes available, and can be used to add the audio file of the performance.

- Click Project and Fit to Music to extend the display of the image (and thus the video length) to the length of the audio file.

This simple process creates a video that consists of a static image with musical accompaniment, an effective if rather "flat" presentation of the material.

To improve the aesthetics of the presentation, a few modifications were made to the above procedure. An animation, such as a pan, zoom, fade, or rotation, may be added to the image by selecting one from the Animations menu. With only a single image comprising the video portion, the animation may result in a very slow, and still boring, visual effect. To enhance this, the cover image was added several times in the first step. This spreads out multiple image "frames" across the length of the audio track. Applying animations to each instance of the image creates faster movement and shifting visual interest. Fade transitions can be added between each instance of the image for smoother transitions between animations.

In a final aesthetic step, titles and captions were added near the bottom of the video. These identify the piece and attribute it to our collection in case the video gets embedded in a website other than YouTube. Both of these functions are located on the Home tab in Windows Live Movie Maker.

Step 3: Uploading the Videos to the Account

Windows Live Movie Maker prominently features a YouTube Share button on the Home tab. However, its use requires signing into a Microsoft Windows Live account. Instead, the following process was used:

- Click the File menu and Save Movie.
- Choose For High-Definition Display under Common Settings.
- Ensure Save As Type is "MPEG4/H.264 Video File (*.mp4)" and choose a file name and location.
- Upload the .mp4 file through the Video Manager interface on the YouTube website.

Step 4: Determining the Appropriate Privacy Level

Once a video is uploaded, content owners can specify the initial privacy level:

- private: visible to the account that uploaded it and other Google accounts specifically granted access to that particular video
- unlisted: not retrieved by searches or included in suggestions, but may be seen by anyone who has the link

- public: fully indexed and searchable, viewable by anyone, and may appear in search results or related videos suggestions

When uploading videos, the authors suggest setting the initial privacy setting to "private" so that the video's metadata can be configured before it is exposed to the public.

Step 5: Adding Metadata

Metadata for a YouTube video consists primarily of the title, description, category, and tags. In choosing the title and description for the videos, care was taken to be both descriptive and mindful of the effect the chosen data will have on the visibility of the content. In the description field, the web address of the corresponding digital collection page was placed in the third line. The YouTube interface only shows the first three lines of the description without clicking Show More, and those first few lines may be all that most users see. YouTube tags provide the rest of the data that affects search indexing. These were chosen based on broad content descriptors and the suggestion of the YouTube software. Tags are not generally visible to end-users but will link the video to highly relevant content elsewhere on the site and increase the chance that someone will find and watch it.

Step 6: Creating a Playlist

Finally, to link the separate videos together, a YouTube playlist was created with a description of the collection. Each of the ten videos was then added to the playlist. This ensures that when a user views one of the videos, a link to the next video in the playlist is displayed. Playlist files also make particular sense for music, since the songs play back to back automatically like an album.

CONCLUSION

Other public and academic libraries with historic sheet music collections can do a similar project. It takes a collaborative effort to accomplish a YouTube project such as this, however, and it is best done in stages. Patience, persistence, and commitment are necessary ingredients for success. So is the combination of skilled people and the equipment to do the project well. The skills and talents of the library's employees should be considered first. Perhaps a joint project can be done between libraries to pool resources if key people or equipment are lacking. The support of library administration is also vital so that appropriate equipment is available to digitize and create content in various formats and media.

It is gratifying to know that pieces from the Gordon Colket Sheet Music Collection are being viewed and used. Reports show that people around the globe have viewed the YouTube clips. More pieces will be performed and added to the digital collection and the YouTube playlist. Ultimately, the Colket Sheet Music Collection stored in preservation boxes no longer lies dormant. It is readily available for others to perform and enjoy.

REFERENCES

MacAyeal, Greg. 2014. "Making Music Collections Come Alive." In *Bringing the Arts into the Library*, edited by Carol Smallwood, 97–106. Chicago: ALA Editions.

SIUE (Southern Illinois University Edwardsville). 2014. "Colket Illustrated Sheet Music (Southern Illinois University Edwardsville)." CARLI Digital Collections. Accessed March 28, 2014. http://collections.carli.illinois.edu/cdm4/index_sie_cwmusic.php?CISOROOT=/sie_cwmusic .

"Terms of Service." 2010. *YouTube*. https://www.youtube.com/t/terms.

Chapter Three

Filtering Google Search Results Using Top-Level Domains

John H. Sandy

When searching Google, users hope to discover appropriate information with minimal effort. A few keywords are entered in the Google search box, and search results are quickly returned. If following this approach, and the goal of a search is to find the best sources, the results may be less than expected. However, when search results are filtered by particular top-level domains, other than .com, high-quality content is readily displayed in the top pages. Depending on the top-level domain used, content retrieved may differ, but it is nonetheless pertinent.

After search terms are put in the Google search box and entered into the system, Google uses a sophisticated algorithm to process terms and find content that is then returned back to the searcher. Relevancy of search results is determined by over two hundred factors. PageRank is one of the factors considered. In the course of processing a query, the Google algorithm gives "credit" to possible importance of a website by factoring in the number and quality of sites that link back to it (Google 2014). Basically, a website that is frequently linked to by other sites may get a higher ranking in search results. In a broader sense, the factors used in the Google search algorithm are proprietary information. Some search experts have tried to decipher what goes into the algorithm by evaluating what is returned from various search queries, but this is mainly guess work.

How to get the best possible search results can still be a challenge. A central issue is user behavior. Due to a variety of reasons, ranging from low attention span, lack of time, technology issues, or other matters, users typically view only the first couple of pages of search results, or, at most, scroll down a few additional pages. If only some interesting results are sought, then any information may meet needs. But if one assumes that the purpose of the search is to find the best content possible, a different search strategy that may improve relevancy is worth pursuing.

With Google search, users have two main options:

- The first is to choose the simple search box on Google.com
- The second choice is to use Advanced Search, accomplished by selecting a special setting from the Google home page

The vast majority of searchers likely settle on the first option. However, with both options, users can choose a precise and productive way to find quality websites. This entails use of the domain name as part of the search string. This approach is of special interest to many sophisticated searchers.

ABOUT DOMAIN NAMES

The domain name offers users a way to capture information coming from special areas of the web and essentially filter out the rest. The domain name is part of a naming convention, used in Internet addresses, a piece of which is the top-level domain (TLD). Commonly recognized generic TLDs are .com, .org., .net, info, and .biz. Other special TLDs, such as .edu and .gov and various country code domains, such as .us (United States) and .ca (Canada) are prominent in Internet addresses as well.

Registrants of domains must frequently meet special requirements when securing a name. Use of .edu is restricted to postsecondary institutions, for example. Initially, .us was open only to state and local governmental organizations, including public schools, with a location in the United States. Later the restriction was removed, and anyone in the United States can now register a .us domain. The .us domain still is widely used by state and local government entities, even while some have moved websites to the .gov domain. The United States government is the major publisher in the .gov domain. The .gov domain was initially restricted to the government of the United States, but this, too, has opened up a little with state governments now able to host .gov sites. In the case of .org, this space was initially intended for nonprofit organizations. But over time, due to lack of control by issuing organizations, .org became widely used by other publishers.

The number of active domains on the Internet is an indicator of the number of websites published in a given domain. The largest volume of domains is found within the .com area. When the domain system was first set up, .com was the designated place for commerce. Today .com maintains a lofty position with over 113 million registrations. Still, even with the massive scramble to land in .com space, other TLDs are an important place to find special content in agreement with the purpose for which they were first created. Being less common than .com but of much value, they are: .org with 10.5 million domains; .us with 1.7 million domains; .gov with 5,804 domains; and .edu with 8,553 domains (DomainTools 2014). While eligibility to register these domains has changed, for some more than others, over the years, searchers can exploit these domains for good purpose by using them to filter search results.

DOMAIN NAME FILTERING

Indeed, the power of domains as filters shows considerable promise when used to locate quality scientific, technological, and medical information. In the United States significant research is done by federal agencies, such as the National Aeronautics and Space Administration (NASA), the United States Department of Energy, and the United States Department of Agriculture, to name a few. Content from these departments is hosted by web servers found on the Internet under .gov. Valuable work on energy, done by the National Renewable Energy Laboratory, is hosted at .gov, for example. And the National Aeronautics and Space Administration, also .gov, reports on discoveries of planets in distant areas of the Milky Way galaxy.

Table 3.1 illustrates search results on emergence of nanoscience filtered by .gov. As might be expected, search results are quite good. Web publications of the United States government are well vetted before online publication and are written and prepared by specialists. Hence, the content is considered accurate and authoritative. Significantly, publications in the .gov domain report on areas of ongoing importance and the latest information.

Similarly, state and local governments produce a wealth of research and compile massive amounts of data and information on topics such as forestry,

Table 3.1. .gov TLD—Sample Search

Search query: site: .gov emergence of nanoscience

Top Search Results

- Toward the Emergence of Nanoneurosurgery
- Nanoscience Research for Energy Needs
- Theory and Modeling in Nanoscience
- Re-Engineering Basic and Clinical Research
- X-rays and Neutrons: Essential Tools for Nanoscience
- About Nanotechnology—Nanotechnology Research
- Nanoscience—US Embassy
- An Analysis of the Educational Significance of Nanoscience
- CNM Overview Brochure
- Nanoscience at Brookhaven National Laboratory

wildlife, aquatic resources, and much more. This content is offered for free and is hosted on web servers operating in .us space or now, on sites quite often, in .gov. For information on the mountain pine beetle infestations in the northern Rockies, a search filtered by .us, retrieves key information from state agencies in Montana, Oregon, and Idaho, for example. Since this pest knows few boundaries, interesting information from localities such as Helena and Missoula, Montana, also shows up in search results conducted in .us space.

Table 3.2 shows results retrieved in a search for information on the mountain pine beetle. By filtering a search on the mountain pine beetle in the .us domain, highly relevant research and related information is found. Of special importance, studies done by scientists and others who work in the areas affected by the mountain pine beetle are easily discovered. In addition, due to the way .us is structured, websites of city and county governments are also retrieved. Urban forests are affected by the mountain pine beetle as well as large tracts of national forestlands. In a way similar to .gov government entities, state and local government publishers operating in the .us domain are a source of timely and authoritative data and information.

In the .org TLD, scientific, technological, and medical information, while usually not research as such, is still very valuable for solid information and understanding progress and policy. Rich data is often a bonus found on .org sites that have a scientific focus.

If a searcher is interested in wildlife, considerable information on mammals is found on the site of the National Wildlife Association, in .org space, for example. Essential information of fisheries management of the Pacific fishery is found on a site hosted by the Pacific Fishery Management Council, also in .org space. Help with issues related to the environment is hosted on the Internet by the Piedmont Environmental Council, hosted under .org.

Table 3.2. .us TLD—Sample Search

Search query: site: .us mountain pine beetle

Top Search Results

- Mountain Pine Beetle—*Wikipedia*
- Mountain Pine Beetles—U.S. Forest Service
- Rocky Mountain Bark Beetle
- Link to Mountain Pine Bark Beetle Information
- Town of Estes Park, Colorado—Pine Beetle Mitigation
- Buford the Mountain Pine Beetle
- Product Use to Prevent Mountain Pine Beetle
- Mountain Pine Beetle
- Mountain Pine Beetle Infestation: City of Helena
- Mountain Pine Beetle—Jefferson County, Colorado

Many types of searches will benefit when filtered with .org. If searching for health or medical information for self-help or to gain another insight from information provided, .org websites produce good results. Major organizations, such as the American Diabetes Association and the American Heart Association, publish websites in .org space. As with .gov and .us, much information in .org space is well vetted and authoritative. The scope of publishing in .org spans almost everything of potential interest, including such areas as environment, technology, medicine, and science. In table 3.3, a search conducted on treatment for diabetes in .org produces important and authoritative information. Again, the filer by TLD drives good content to the top of search results.

The .edu domain is a special case, since it is focused mainly on postsecondary education organizations. Colleges and universities operate in the .edu space. These sites host a wide variety of information. If a searcher wants content for study or learning on scientific, technological, or medical topics, .edu space is especially useful. It is noteworthy that, in the era of open access, many scholarly preprints are hosted on servers in .edu space. Educational institutions have a wide variety of interests. As such, content on a wide variety of general and administrative content is present and may intersperse with technical information in search results. Nevertheless, technical terms used in a search query when matched by Google with terms in websites will generate good search results.

Search queries done with .edu illustrate the quality of information that can be found using this approach. A search on global warming filtered by .edu finds an abundance of quality information from the University of Washington and forecast data simulations done at the University of Chicago. Major institutions of higher education in the United States are heavily engaged in research, much of which is funded by the U.S. National Science Foundation and various high-profile foundations, making .edu space a rich trove of tech-

Table 3.3. .org TLD—Sample Search

Search query: site:.org treatments for diabetes

Top Search Results

- Treatment & Care for Diabetes
- Type 2 Diabetes
- Diabetes Mellitus Treatment
- Spotlight on New Diabetes Treatments
- Type 2 Diabetes: Symptoms, Causes, Treatment, and More
- Treating Type 2 Diabetes
- Treating Type 1 Diabetes
- Diabetes Management
- Traditional Plant Medicines as Treatments for Diabetes
- Prevention & Treatment of Diabetes

nical information. Work done by students and others in areas related to instruction and administration is found in .edu space as well. If the purpose of a search is to locate technical information, evaluation of authorship and sponsoring entity in .edu space is a must.

In table 3.4, a search for information on global warming generates much significant content. Since institutions of education are engaged in advanced research, the value of information obtained in the .edu domain is very apparent, in this example.

In some cases, an alternative domain search strategy employing metasearch technique may produce an appropriate mix of resources from a combination of .gov, .us, .org, and .edu space. In the Google search box desired keywords are entered along with a sequence of domain filters combined with the Boolean operator OR. For example, a full search string might be entered as: *treatment diabetes site:.us OR site:.gov OR site:.org OR site:.edu*. While quite powerful and potentially useful, this method diminishes immediate visibility of content from any single domain.

FILTERING FOR IMAGES AND VIDEOS

In addition to use in general search, TLDs are excellent filters when used with special formats. Beyond website retrieval, Google search is useful for finding images and videos. Both images and videos can be filtered by .gov, .us, .org, and .edu in much the same way as done with websites. An examination of search results for the mountain pine beetle in the .us domain produces a bounty of images. The images are mostly photographs, but maps, charts, and tables are displayed as well.

Google controls for display of images when filtering with .us appear a little less than perfect, since a couple of images from .org and .gov are

Table 3.4. .edu TLD—Sample Search

Search query: site:.edu global warming

Top Search Results

- Global Warming Effects Map
- Home, Climate Change, US EPA
- A Students' Guide to Global Climate Change
- Global Warming News
- Virtual Courseware Global Warming
- Video Lectures: Global Warming
- Models—Global Warming: Understanding the Forecast
- Global Warming—Research Issues
- Global Warming in North America
- Yale Program on Climate Change Communication

displayed alongside images from .us space. Nearly all the images are from .us, however. Again Google retrieves images from websites published by state, local, and other government agencies. Interestingly, in this example a single image of damaged forest land is retrieved from a website published by a public school in Colorado. Public schools publish in the .us domain.

Filtering for images is a good choice in .org space, as well. In the case of a search on treatment of diabetes, a wide and rich assortment of images is retrieved. Images cover topics ranging from a diagram of an algorithm for glycemic control to a table listing oral agents for treatment of diabetes. Several images show apparatus and methods for collecting blood.

When filtered for images in .edu space, images for global warming show abundant data and information in tables, graphs, and maps. Many photographs are of landscapes and nature, which may experience the aftermath of global warming. Since .edu publishes education-related materials, many images are hosted in faculty blogs.

The National Aeronautics and Space Administration (NASA) and several of its sub-agencies publish photographs and other images in .gov space. Google searches for images in .gov lead directly to work of the United States government. The images in .gov are about subjects that are beneficial for study, research, or just to be informed about. In a search for Earth-like planets, use of the keyword *Kepler* along with the term *exoplanets* would perform precisely and push Kepler mission images to the top. Kepler is the specific mission sent out by NASA to probe for Earth-like planets in space beyond the solar system. The search results from this approach to filtering are immediately on target.

Search results when filtering videos using .us, .gov, .org, and .edu domains may, in some cases, deliver less content. In the search on *mountain pine beetle* few videos are retrieved. This is likely a function of the subject matter and does not suggest anything about the capability of Google to filter content retrieved using the tab for videos.

GOOGLE ADVANCED SETTING FOR FILTERING

It is unlikely that most users are familiar with the domain system and how it can be used to improve search results. The command driven format, as shown in the tables in this chapter, may be more than most searchers can understand or will even attempt to learn about. Google seeks to bridge the gap, somewhat, with its advanced feature. Unfortunately, the advanced feature, which presents a multistep dialogue page for users to enter search terms/parameters in a selection of predefined "fields" is somewhat hidden in the lower right-hand corner under a link called "Settings."

A user so inclined can click on the Settings link and when a text box opens choose "Advanced Search." The advanced search page under a section called "Then Narrow Your Results By" gives searchers the ability to specify site or domain. An explanation of what this means is shown in a message near the entry box, saying, "Search one site (like Wikipedia.org) or limit your results to a domain like .edu, .org, or .gov." However, the full implications of taking such action are not evident to those who are unfamiliar with what's being suggested. Still, the domain search approach is offered inside Google and refine by domain is easy and simple to do.

NEWER TLDS AS FILTERS

Domain names are always changing. As recent as 2014, the Internet Corporation for Assigned Names and Numbers (ICANN) introduced numerous new TLDs making name space even more robust. ICANN reported that 175 new generic top-level domains have been delegated (ICANN 2014). Among the new TLDs, domains such as .engineering, .technology, and .computer may hold considerable promise when used as filters when searching in these areas. A few years earlier ICANN delegated space for .museum (reserved for museums, museum associations, and individual members of the museum profession) and .aero (reserved for aviation), domains also potentially useful as filters.

Both .aero and .museum effectively and efficiently pull specialized content. By using these domains as filters, users are assured of quality content. A search for Cessna along with .aero guides users to an abundance of information on the aircraft. Since museums serve an educational mission, searching for minerals and .museum delivers considerable value for understanding all sorts of treasured mineral specimens.

USE OF INTERNATIONAL DOMAIN NAMES

There are about two hundred forty international country code top-level domains (*Wikipedia* 2014). Country code domains such as .de (Germany), .it (Italy), and .ru (Russian Federation) are widely used on the Internet and familiar to most. The value of country codes as filters depends on an understanding of the rules under which they are governed by. Many country code TLDs are unrestricted and available for use by anyone around the world. Anyone can register a domain name in the country code .ms (Montserrat), for example. However, in other countries such as the United Kingdom, Australia, and China, domains are carefully structured and controlled. A technique using special second-level domains, immediately preceding the top-level domain, is used by some countries to segment Internet space. In the United

Kingdom, the domain name convention is .ac.uk for institutions of higher education, .gov.uk for government entities, and nhs.uk for national health services. In China a similar convention is followed with .gov.cn for government, ac.cn for academic institutions, and .org.cn for nonprofit organizations. A full investigation of how a particular country code domain is structured and administered is usually an important step to take before using it as a filter.

SUMMARY

The ideas and processes presented in this chapter provide valuable insight for getting more out of Google search. Google search in and of itself is a powerful search engine. Its impact on finding information in all spheres of human inquiry is legendary. In its quest to keep search as simple as possible for nearly everyone, Google search, in its basic search box, underutilizes some potential. Searchers may not always find all relevant content if they only use the basic search box to interact with the system. For many users, the results obtained from the basic search box may be all that's needed. However, it's important to understand that results retrieved from the basic search box are an amalgam, driven by the Google algorithm. Some high-quality websites and premium scholarship may thus go unnoticed. Content may be hidden much lower in the search results, not in top pages. By employing the power of domain names for filtering content, relevance may be enhanced to some degree. Certainly the efficiency of the process is improved, since domain filtering, almost by definition, leads to organizations that produce much of the world's best content on the web.

REFERENCES

DomainTools. 2014. "Domain Count Statistics for TLDs." Accessed April 14, 2014. http://www.domaintools.com/statistics/tld-counts/.

Google. 2014. "How Google Search Works." Accessed April 14, 2014. https://support.google.com/webmasters/answer/70897?hl=en#%20/.

ICANN. 2014. 175+ New gTLDs Delegated. Accessed April 14, 2014. https://www.icann.org/en/news/announcements/announcement-24mar14-en.htm.

Wikipedia. 2014. List of Internet Top-Level Domains. Accessed April, 16, 2014. http://en.wikipedia.org/wiki/List_of_Internet_top-level_domains.

Chapter Four

Google Digital Literacy Instruction

Richmond Public Library

Natalie Draper

The Richmond Public Library in Richmond, Virginia, has used free Google applications to facilitate a scalable digital literacy program, allowing us to get the most out of a very limited technology budget while responding to the increasing technological needs of our patrons. Faced with outdated equipment and software and limited staff we adopted the Google Chrome browser and the Google suite of applications to design an array of digital literacy classes and improve the user experience in our computer labs.

Many patrons we assist depend entirely on the library for access to computers and the Internet. Depending on public computers means that they relied upon extra moving parts like USB drives to store, manage, and transport their files. I've seen many patrons struggling to use their e-mail accounts for file storage only to have their careful formatting lost when they attempt to open the file with a different version of Windows on a different computer. USB drives are hardly secure (Pagliery 2014), and who doesn't have a lost-and-found box full of those little drives behind the circulation desk? After observing many such situations in the computer lab I developed an introductory computer class based on creating and managing files "in the cloud" using Google Drive. This way, regardless of operating system, availability of Microsoft applications, and independent of accessories, our patrons can safely access, create, and manage their files from any computer with access to the Internet. Since the cloud is likely to be around for a while, getting students started with emerging technologies is a forward-thinking approach to digital literacy.

One of the great things about the Google suite of applications, and how it can benefit digital literacy instruction, is how responsive the features are. For the novice user Google Drive appears to be rather spare; it doesn't overwhelm with every available feature on display. Even paragraph styles and bulleted lists are hidden behind the "more" button. This presents the user with a pared-down, icon-driven row of basic formatting options, and a slim line of drop-down menus that resembles an older version of Microsoft Word.

Far more than just a search engine, the Google suite of applications includes familiar services such as Google Maps, News, and Image Search. For social media, content management, cloud storage, and media sharing Google offers Blogger, YouTube, Picasa, Plus, and Drive (formerly Docs). Introducing users to the free and versatile Google iteration of familiar programs gives them a good start on up-to-date computer literacy.

The following two lesson plans, Basic Computing with Google and Research and Cite Like a Pro, were designed and implemented at the Ginter Park Branch of the Richmond Public Library. Our equipment needs were minimal; six laptops were sent via courier from the main library for twice-monthly computer classes, and we set up a laptop connected to a projector and a projector screen. Chrome was installed on every computer and set as the default browser. Patrons were required to register in advance of the class and provide us with a phone number so that we could call to remind them of their appointment on the day before class met. We created a Google account for the branch to use for instruction purposes only. With this account we are able to manage lesson plans, access and deliver presentation materials, create surveys, and manage survey data, all under one login from any computer. The cloud makes it simple to collaborate and share materials with colleagues across the system.

BASIC COMPUTING WITH GOOGLE

This class has students set up a Google account and learn the basics of how to search the web and e-mail safely and securely from any computer. The class was developed for seniors with little to no experience using computers. It will take the student from launching a browser icon to creating a Google account and sending an e-mail to conducting a search for information using Google. This class can be divided easily into two sessions. When using the projector to demonstrate key points, ask that the students take their hands off the mouse and keep their eyes on the screen. Eager students will attempt the exercises before the rest of the class is ready, causing a sort of instructional bottleneck.

Browsers

Analogies can be helpful in illustrating abstract and unfamiliar concepts. A freeway analogy proved effective and drew some chuckles when explaining the difference between the Internet and a browser. The Internet is the free-way; a browser is your car. All browsers take you to the same Internet, just as all cars can get to the same destination. A Plymouth gets you to Kroger just the same as a Maserati, right? But one might be faster or more reliable, one might be safer, one more aesthetically pleasing, and so on. For this class we will be learning how to drive Google Chrome. If you have any experience using Internet Explorer (the blue E) you might notice that Chrome "handles" a bit differently, but you might also appreciate the speed and simplicity of Chrome.

Show the icons that represent the most commonly used browsers (Chrome, Internet Explorer, Firefox, and Safari) and explain where they will find them, how they will look different, and which you recommend they use. We recommend Chrome on Richmond Public Library computers and use it in our classes. Have the students practice double-clicking the icons to launch their browser.

Setting Up a Google Account and Sending an E-mail

In most cases Google requires a working cell phone to create an account. Encourage students to bring a cell phone with texting capabilities to class. It is possible to use the Gmail account created for the library or branch as a backup e-mail in case using a cell phone is not possible. I have seen students share phones and have even used my own phone (not recommended) in order to set up the initial accounts.

Discuss the requirements for usernames and passwords and be clear about how they must pick a password they can remember but others can't easily crack. "Password" is a lousy password. Google passwords require a minimum of eight characters. Encourage the class to use a combination of letters and numbers and emphasize that case is important. Liken passwords to safe combinations.

Demonstrate signing in to Google by signing in to the library instruction account. Invite the students to log in with their newly set up accounts. Advise the students to uncheck the box next to "keep me logged in," especially when using public computers. Incorporate this lesson as a step to signing in rather than treating it as an aside. Safety should be stressed throughout this lesson. Online account providers want to keep users logged in at all times. Stress how important it is to always log out of their accounts, especially on public computers.

Once the students have signed in, introduce them to the key parts of their e-mail dashboard. Have the students compose a message to the library's Gmail address. Require that they include a subject line and type "Testing, 1 2 3" into the body of the e-mail. Have them attach a photo stored on the hard drive to the e-mail. Most computers come with sample images in the pictures folder. Instruct them to click "attach file," "browse," then locate the folder containing images and select the desired image to upload. When you receive their messages, quickly send a reply. Once the reply is sent they will have one unread message in their inbox. Instruct the class to open your e-mailed reply and forward it to a classmate.

Use this opportunity to discuss how to be careful about which e-mails you open, and which to ignore. Stop Think Connect, a coordinated effort of for-profit companies, nonprofits, and the department of Homeland Security, advises Internet users to be cautious about anything that "implores you to act immediately, offers something that sounds too good to be true, or asks for personal information" (Stop Think Connect 2014). I would add that no one with good intentions will ever ask for bank account information in an e-mail. If you wouldn't give that information to a stranger walking by, don't give it to a spammer either. Go on to explain spam and phishing scams, and how e-mail can contain harmful viruses. The STC website offers posters and handouts about staying safe online.

Advertising vs. Content

Pull up an article on any ad-supported website. I use a local network news affiliate's website as an example. Point to different parts of the page and ask students to identify the text, links to other articles, the ads, and links to other relevant content on the site. Discuss "click baiting." Click baiting is a kind of bait and switch where a headline is crafted to entice people into clicking it. That is not to say that it is malicious. The key is to recognize the difference between paid advertising, informative content, and all that lies between (Burriss 2014). Links to "one simple trick," "new rules," and "magic diet secrets" using provocative images abound and are cleverly inserted into web pages to look like links to articles. Advertisers have been using similar tactics for ages in print. Show the students any page from a newspaper with similarly styled advertising.

Searching Google

First, identify the search bar and conduct a sample search. Search for *cookie recipe*. In .28 seconds that search term returned 31,500,000 results. That's a lot of cookies. Explain the parts of a search results page and identify the ads. Ads in Google appear at the top and right of a results page, and are indicated by a yellow badge that says "ad." Suggest ways to narrow down a search. Typing *cookie recipe site:smittenkitchen.com* will retrieve only results within Smitten Kitchen.

Besides just searching for websites, Google can provide quick answers. Metric to English conversion, calculations, local weather, and currency conversions are examples of quick answers Google will fetch and display at the top of a search results page (Google 2014). Demonstrate the calculation feature in Google search. Identify the search box again. Enter any equation into the search box, using +, -, *, or / for addition, subtraction, multiplication, and division, respectively, and press enter. The equation will display in a calculator on the results page.

Instruct students to "ask" Google how the local weather is by typing the word *weather* and the name of your city in the search box and pressing enter. Invite students to "ask" Google how many ounces are in a pound or how many liters make up a gallon. Demonstrate how to find the dollar value of the yuan. Beginners can become easily frustrated and overwhelmed. Demonstrating some of the "wow" features in Google can lighten the mood.

Assessment and Survey

Create a survey in Google Forms. Surveys can be set up to require clicking to fill in a circle when only one item in a list can be selected or to check multiple boxes when asked to "check all that apply." Text can be typed into blank fields to provide a short answer, and a button must be clicked to submit the survey. I designed a survey that asked students to check all the reasons they were taking the class, select their age group (only one circle could be checked), and type a short answer to a question about how they liked the class and what could be done to improve it. Once they click the "submit" button their surveys are automatically compiled into a spreadsheet.

Case Example

John Doe, an eighty-five-year-old auto mechanic with no computer skills, would like to have an e-mail account and create and manage an online presence for his business. One class is not enough to give Mr. Doe all the support he needs, but it gets him started. He returns to the library computer lab every few days to log in to his new Google account to work on his documents. Library patron Mr. Doe was the inspiration for Getting the Most out of Google. He is eager to learn and continues to come back to use the computers and ask questions. Drive allows him to log in to any computer at any location and pick up right where he left off without fear of losing his work, corrupting files, or misplacing a flash drive. He need only remember one username and password to manage his e-mail, documents, and page for his business. He doesn't have to worry that he didn't save his work because Drive automatically saves as you type. Mr. Doe has become much more secure and independent when it comes to computers and the Internet.

RESEARCH AND CITE LIKE A PRO!

By the end of this course students will be able to conduct a Google search for information and to critically evaluate the results for inclusion in a research assignment. This lesson plan is designed for middle and high school patrons.

Many K–12 schools are adopting Chromebooks for student use (Reid and Strong 2013). They are inexpensive and simple to use, but rely upon an Internet connection and use Google applications. We now have a generation growing up using Drive as their main word processor, and Google as their primary research tool. For students, a significant advantage to using Chromebooks and the Google suite of applications is the ability to securely access their documents, e-mail, search history, and bookmarks from any computer, regardless of operating system, version of Windows, or other specific software. Students lacking access to a computer in the home will find fewer obstacles to saving, storing, and retrieving their documents. Our biggest challenge in digital literacy is to give students the tools to find, filter, and evaluate the information they need in a vast sea of constant input.

Set Up a Document in Drive

Open a new document in Google Drive. Name the document and begin typing a thesis statement. Highlight a search term and right-click the highlighted text. Select Research [term] from the menu. This will bring up the research pane on the right of the document display. A *Wikipedia* article is often the first search result.

Refining Search Strategy for Better Results

Google allows users to

- restrict the search to a specific website or type of site (domain), such as government, educational, or nonprofit websites (.gov, .edu, .org, respectively) by adding *site*, a colon, and the relevant website or domain after the search term: *civil war site: www.va.gov* or *national parks site: .gov*;
- "fill in the blank" or do a truncated query, indicated by an asterisk (*): typing *librar** will bring results including libraries, librarian, and library;
- search for either of two or more words using OR. For example, *Libraries OR Librarian OR Library*;
- find information about a site by typing *info*: before the website name, for example, *info:Google.com* to retrieve information about the website;
- exclude a word or site from search results with a minus sign: *-librarians* or *-site:wikipedia.org*;

- search for an exact word or phrase using quotes: *"she walks in beauty like the night"* will retrieve pages containing that exact phrase; and
- search for pages that are similar to a website; for example, type *related:nytimes.com* to retrieve other newspapers.

How to Evaluate Sites

Check for authority, accuracy, objectivity, currency, and coverage.

Authority

That's who wrote it and their reputation. Look at who the author of your online source is. Is this person a trusted scholar or journalist with credentials to back up his or her authority on the subject at hand? Or is the author some guy with a Twitter account and a lot of opinions?

- Examine different sources side by side. *Wikipedia* is a good start. Discuss what credibility means to the students and how they would assess it.

Accuracy

One way to assess accuracy is to look for a website's guidelines for submissions. Even though it has often been maligned by teachers for unreliability, *Wikipedia* does require that articles submitted "should cite reliable third-party published sources . . . credible published materials with a reliable publication process [and] authors [who] are generally regarded as trustworthy or authoritative" (*Wikipedia* 2014) according to their guidelines.

Currency

This just means how recent the information is, the importance of which varies from topic to topic. Was that article about new technology written in 1998? Ask yourself: Is the site currently being maintained? How would you know this? If the links are updated and working properly is one indicator. Google typically returns the most current relevant links first, but sometimes results can be quite old. While an old article about the battle of Gettysburg is probably fine to use, what do you think a teacher would say about an article on "new" Internet technology that was written in 2001?

- For an exercise try doing an advanced search using Google News. The advanced search is indicated by a downward arrow to the left of the blue Search icon. Click it and demonstrate examples of how to refine a search.

Objectivity

Is that blog post written in all caps? Information can be truthful, though not necessarily objective. Accuracy is one factor in determining objectivity, but impartiality also plays a role. The personal account of a crime victim is accurate, but hardly objective. An eyewitness testimony might be impartial but could lack detail. A news story that includes both eyewitness and victim accounts would be considered objective.

Coverage

Does the information appear to be complete and comprehensive?

- Examine a *Wikipedia* article (Google's top result, for the evaluation criteria. Does it satisfy our requirements for accuracy? Authority? Objectivity? Currency? Coverage? Examine something that would get a lot of traffic and therefore probably be heavily moderated. Examine something "locked" by editors that might be controversial such as the Haymarket Affair (Messer-Kruse 2012). *Wikipedia* includes a "talk" page for debates over article revisions.
- Show a sample search screenshot of a major topic. I use the results page for a simple search: *Virginia*. Direct the students' attention to the synopsis in the right panel. It's convenient, isn't it? Note where the information comes from, though. See the source indicated right after the summary? *Wikipedia*! Watch out for convenience. Google's search results page will now include a summary, often drawn from *Wikipedia*, to provide searchers with a quick "answer."

Citation

Sources can be cited immediately from the research pane. Remind the students of the importance of citing their sources. Cutting and pasting from the Internet is plagiarism, plain and simple. To select a format for citations, click the drop-down arrow below the search box in the research pane. MLA, APA, and Chicago citation styles are offered.

Collaboration

Every student is assigned group projects at some point in school so the ease of sharing and collaborating on documents makes for an engaging lesson for any age in a computer class. Share a document created in Drive with the entire class by clicking the blue "Share" button at the top of the page. Enter the Gmail addresses of the students and invite them to view, comment, and make changes to a group project. Consider creating a mock *Wikipedia* article about the library or branch for the students to collaborate on vetting.

OTHER DIGITAL LITERACY RESOURCES

Google Plus and Maker Camp

Libraries can register as campsites for the Maker Summer Camp, a "free summer camp from Google and Maker Camp for building, tinkering and exploring" (Maker Camp 2014). The camp meets for several weeks each summer via Google Hangouts, a video chat service provided by Google. The programs, designed for tweens and teens, are geared toward science, technology, engineering, and math activities. There are weekly projects to build and many opportunities to participate in Q&A sessions with tech professionals and collaborate virtually with young people around the world.

Build with Chrome

When it comes to programming for children and young adults, it doesn't get much cooler, or freer, than Build with Chrome, a partnership between Google Chrome and Lego ("Build with Chrome" 2014). A free Chrome app that actually works in any browser (but not mobile yet), Build with Chrome resembles a simplified 3D modeling program. Builders use keyboard commands and virtual Legos to create whatever they can imagine. As an experiment to expand our digital literacy efforts to include young children, we added Build with Chrome to an existing Lego Club program for five- to twelve-year-olds. We connected a Chromebook to the projector so that the children could see their three-dimensional Lego sculptures projected on the wall. The results were pretty outstanding.

The Goodwill Community Foundation Free Online Learning

The Goodwill Community Foundation is a great place to direct students for additional lessons. "From Microsoft Office and e-mail to reading, math, and more, GCFLearnFree.org offers 125 tutorials, including more than 1,100 lessons, videos, and interactives, completely free" (Goodwill Community Foundation 2014). GCFLearnFree offers several Google-related courses including Gmail, Chrome, and "Using the Cloud," Google Drive, Accounts, and more.

Mousercises (Palm Beach County Library)

Brought to you by the Palm Beach County Library System, Mousercises are an indispensable tool for acclimating older students to using a mouse. Brilliant in its simplicity, the program works much like a computer game in which a single player follows instructions to click certain targets gradually increasing in complexity.

REFERENCES

"Build with Chrome." 2014. www.buildwithchrome.com/.

Burriss, Larry. 2014. "Clickbait Not Illegal, but It Sure Is Annoying." *Murfreesboropost.com*. Aug. 3. www.murfreesboropost.com/burriss-clickbait-not-illegal-but-it-sure-is-annoying-cms-40078 .

Goodwill Community Foundation. 2014. "Free Technology Training." www.gcflearnfree.org/.

Google. 2014. "Inside Search: Search Tips & Tricks." www.google.com/insidesearch/tipstricks/ .

Maker Camp. 2014. "Base Camp." http://makercamp.com/.

Messer-Kruse, Timothy. 2012. "The 'Undue Weight' of Truth on *Wikipedia*." *Chronicle of Higher Education*, February 12. http://chronicle.com/article/The-Undue-Weight-of-Truth-on/130704/.

Pagliery, Jose. 2014. "Stop Sharing USB Flash Drives—Right Now." *Money.CNN.com*. August 1–5. http://money.cnn.com/2014/08/01/technology/security/usb-hack/ .

Palm Beach County Library. 2014. Mousing Around: Mousercise! www.pbclibrary.org/mousing/mousercise.htm .

Reid, Zachary, and Ted Strong. 2013. "Laptops Coming to Chesterfield Middle, High Schools." *Richmond Times-Dispatch*, November 30. www.timesdispatch.com/news/local/education/laptops-coming-to-chesterfield-middle-high-schools/article_e87eaf18-5901-11e3-86eb-0019bb30f31a.html .

Stop Think Connect. 2014. "Tips & Advice Overview." http://stopthinkconnect.org/tips-and-advice/.

Wikipedia. 2014. "Identifying Reliable Sources." http://en.wikipedia.org/wiki/Wikipedia:Identifying_reliable_sources .

Chapter Five

Google for Music Research

More Than Play

Rachel E. Scott and Cody Behles

For better and for worse, Google has dramatically changed the music research landscape. Both the proliferation of musical resources in various formats online and the ease of retrieval that Google affords are unprecedented. Google products, including Google Play Music, YouTube, Google Books, and Google Scholar have created a dynamic and immersive environment that facilitates music discovery, research, and performance. These tools can be accessed by anyone with an Internet-ready device and an Internet connection. This means that many users are searching for and finding musical materials outside of libraries. Many will find the resources that they need, but others will be frustrated by the small number, or low quality, of freely available online musical materials. Users who are dissatisfied with the musical materials uncovered by a Google search must be convinced that libraries can play a role in their music-information needs.

Libraries continue to select and acquire sound recordings, musical scores and parts, and reference materials that are not freely available online. More important, librarians expertly connect users with information resources. This expertise is especially useful when searching for specialized materials in various formats, which is frequently the case in both popular and classical music research. In order to continue to help users find the musical materials they seek, librarians must familiarize themselves with new search tools and broaden their understanding of appropriate research sources. In short, librarians need to welcome Google in their libraries and leverage its powerful tools.

This chapter will discuss several Google products that can be used in conducting music research, including:

- Google+
- Google Books
- Google Play Music
- Google Maps
- Google Scholar
- Google Translate
- YouTube

Although some librarians might be uncomfortable with the ways in which Google has altered the research process—specifically, the loss of a controlled and structured search environment—librarians are not obsolete in a world of information overload and keyword searching. How can librarians insert themselves in the online music research process to encourage users to go beyond the easily accessible sources they find online and engage with musical materials in library collections?

GOOGLE FOR MUSIC CONSUMPTION

Many library patrons will have all of their music-information needs met online. This is okay, because many users have only casual needs related to their music consumption. Most individuals engage with music as consumers by:

- paying for and downloading digital files
- buying physical albums
- buying concert tickets
- paying a subscription fee for streaming products such as Spotify, Songza, or Pandora

According to one industry analyst, the average iTunes user spends $12 per year on music (Dediu 2013). Besides making purchases, other online musical behaviors include searching Google to:

- identify a song or artist
- watch and share videos on YouTube or other video platforms
- identify musicians or music of interest

Google Play Music is but one example of a digital music store and cloud-based, music-player platform: iTunes and Amazon Cloud are two others. These platforms allow users to purchase, download, and stream music and videos to various devices. Google Play Music, however, has unique features that are of interest to the music researcher and music consumer alike. Album and artist metadata from Google Play Music have been aggregated and are visually represented in the Music Timeline, a project of the Big Picture and Music Intelligence research groups at Google. With its breakdown of popular music subgenres and bands over time, the timeline could be useful for librarians providing listener advisory outside of familiar genres. Another product that would be welcome at the reference desk is one that could "name that tune." Google Sound Search enables one's device to "listen" to music heard in the background and provide the song name and a link to purchase the track from the Google Play Music store. This is one tool that would outperform most librarians and should consequently be added to the listener-services arsenal. The only caveats are that Google Play Music requires a personal account and either a device running Android or iOS or an Internet browser.

Subscription music services are increasingly attractive alternatives to music ownership. According to the International Federation of the Phonographic Industry (2014), music subscription services' revenues increased 51 percent in 2013: "Global revenues from subscription and advertising-supported streams now account for 27 per cent of digital revenues, up from 14 per cent in 2011." Google is currently negotiating the acquisition of subscription music service Songza. Songza is noteworthy for its curated track lists based on the listener's mood or activity, for example, "watching the sunset" or "feeling fierce." This acquisition would improve Google Play Music's music-recommendation capabilities. Libraries, too, are increasingly signing up for subscription music services. Music databases like Naxos and Classical Music Library have provided library users with streaming access to jazz and classical tracks for over a decade. Mainstream music services, like Freegal and Hoopla, offer access to most popular genres.

Although purchasing music and selecting a subscription music service is a largely personal decision, there is still room for the librarian in listener advisory. Google provides librarians with several means of making recommendations. A Google search of *similar artist* and a band or musical artist's name can yield very useful results from various sources. If a patron were to say, for example, "I really like Bruno Mars, but I'm sick of listening to *Unorthodox Jukebox*, you might search: *Bruno Mars similar artist* and investigate results from sites like Last.fm, MTV, or AllMusic.

Most streaming music platforms have a built-in recommendation feature. Librarians should be aware of the availability, if not the exact functionality, of these features. Google Play Music has the "recommended for you" feature based primarily on previous purchases. Librarians in public-service positions have long provided listeners' advisory, but some librarians have been inspired to mimic the immediacy of streaming platforms and to provide content with their recommendations. Steven Kemple (2014) recently highlighted an innovative listener-services program at the Cincinnati Public Library in his Music Matters column in *Library Journal*. In that program, librarians select a new CD for users based on their preferences and their chosen level of adventurousness and send it to the patron. Even when a user has only casual music information needs, librarians should evaluate how Google products can be leveraged to address those needs.

GOOGLE FOR MUSIC PERFORMANCE AND PRODUCTIVITY

Some users do not passively consume music; rather, they use musical materials to perform or produce musical compositions. Google Play Music users can purchase several music-themed applications from the store, including:

- tuners
- metronomes
- pitch pipes
- recording applications
- drills and tutorials
- music notation
- transposers
- piano keyboards and other instruments

All of these applications can make one's personal device more useful in musical performance and rehearsal.

The first result of a Google search for a song or title and composer is often a YouTube result. Although most users do not contribute to YouTube, this Google product does provide an easy and popular means of disseminating one's musical performances, compositions, tutorials, and products. The quality of this content has been questioned by Dougan, who "found that faculty and librarians do not entirely share perspectives concerning the quality of YouTube's content, metadata, or copyright concerns" (2014, abstract).

Google+ is another means of sharing one's music, whether by posting YouTube videos or hosting Hangouts on Air to present live concerts. In addition to promoting one's own music, there are several examples of collaboration on Google+. Musical compositions have been co-written by geographically distant collaborators, music instruction and auditions have been held virtually, and other music-related uses will continue to surface should Google+ continue to gain in popularity.

Musical scores, or printed music, remain the primary source for classical music research and are used to learn and perform new music. Google products have influenced not only how users find, but also how they interact with, musical scores. Although Google Books and other mass-digitization projects have scanned millions of scores, the optical music recognition software that enables content searching of music is less developed than the character-recognition software used to process text. Users must know the composer and title to effectively search Google for scores. Single Interface for Music Score Searching and Analysis is a project sponsored by international music scholars and librarians that aims to build an online music document analysis and retrieval system and to unite the several disparate collections of digitized scores online. However, its limited progress means that it does not yet provide much competition to Google searches and products.

Google has invested in research and programs that, although not directly accessible to the public, affect how music is accessed and digested through its services. Chief among these is Waveprint, an audio identification program. Waveprint assigns a unique identifier to audio and video files in the form of a wavelet, or amplitude oscillation, that is resistant to variations and manipulations in audio files, making searches more consistent and accurate (You, Chen, and Chen 2013). Waveprint also allows Google to embed details such as copyright information into files in a way similar to that employed in Google Books, with the added benefit of being able to prevent piracy. This is possible because the wavelet cannot be modified easily, so manipulations of the file can be traced to their original source (Garcia-Hernandez, Feregrino-Uribe, and Cumplido 2013). Waveprint is present in applications such as the Sound Search Application and through the Google Search platform on Android devices and in Google Play.

Google's investment in the Bodleian Library's What's-The-Score Project, a collaborative open-source catalog of score metadata (Johnson 2012), and its acquisitions of Simplify Media in 2010, RightsFlow in 2011, and its pending acquisition of Songza indicate an interest on the part of the company in the development of more dynamic music metadata. Community interest in music information and metadata is reflected in Google+; communities like Sheetmusic Playback of Original Music and Online Music Collaboration show the potential of Google products in the development of rich exchanges of information among musicians.

In the meantime, in order to conduct an exhaustive search of digitized scores, users have to search various platforms individually. Many users, however, prefer a single, simple Google search to find scores (Dougan 2012; Oates 2014; Scott 2013). The International Music Score Library Project (IMSLP) is a large-scale score digitization project and is well indexed by Google. A Google search of a known composition (for example, *Stravinsky Firebird*, or *Mozart An Chloe*) will often yield an IMSLP score on the first page of results.

Some might wonder what is wrong with this arrangement. If users are finding what they search for, why should a librarian care? The problem is that not all musical scores are created equal; the quality of a musical score depends on the editing, printing, quality of digitization, availability of contextual information, and other factors (Scott 2013). IMSLP offers users the opportunity to purchase the score of interest; but unlike Google Books and

Google Scholar, it does not offer users the opportunity to find it in a library. This referral inserts the library into the music-research process and is a great talking point during instruction and the reference interview. Another way to insert the library into a Google search for a score is to ensure that holdings have been set in WorldCat and that integrated library systems and digital repositories have been optimized for Google harvesting. By doing so, the item bibliographic record may appear in a list of Google results.

GOOGLE FOR CLASSICAL MUSIC RESEARCH

As previously mentioned, scores remain the most important source when conducting research in classical music. Secondary sources, however, are of great importance and should not be ignored. Many academic library users are very familiar with Google Scholar and search it first, instead of library subscription databases like JSTOR, IIMP, or Project Muse. Several studies have investigated how comprehensive Google Scholar is relative to other databases or within a given discipline, for example, in medicine (Giustini and Boulos 2013). Google Scholar searches several important music journals, but the coverage is inconsistent and far from thorough. Google Scholar should not be searched as a replacement to specialized music databases, but as a supplement to those databases. Its wide-ranging coverage makes it especially useful for those conducting interdisciplinary research.

A search of Google Scholar should not be discouraged. Those users who are more confident searching Google Scholar can use it to identify quality published materials and then obtain the materials through their library. The Google Scholar platform may be customized according to user preferences by

- adding Library links for all of the libraries to which the patron has access
- selecting the bibliography manager of choice (including BibTeX, EndNote, RefMan, and RefWorks)
- choosing collections of interest (articles, patents, and case law)

On the backend, librarians should contact their link resolver vendor to ensure that their subscriber IPs will be included and that their links appear in Google Scholar.

Google Books contains many titles of interest to music researchers, several of which are available in full view. Several nineteenth-century periodicals, including *Allgemeine musikalische Zeitung*, enrich Google Books' music holdings, as do reference titles like the 1904 and 1918 *Grove's Dictionary of Music and Musicians*. There are, however, limitations on the coverage, both for the periodical runs and the reference sets. Although it is wonderful that users can discover these important reference tools in a search platform that is easy to use and comfortable, it is possible that some users might interpret a lack of presence in Google Books as a lack of coverage. A second disadvantage is that many users must first be taught how to use music-reference tools. If a student were to encounter a thematic catalog—or worse, something labeled *Thematisches Verzeichnis*—she would not likely think the source could provide the location of manuscripts, documents associated with the composition, or the date and location of the first performance.

A further disadvantage to Google Books is the lack of current titles available in full view. Although historical titles are an important component of historical musicology, recent titles in the discipline are essential to rounding out research. These are more often findable, as Google Books searches most major publishers, and may help with the discovery of library resources. This addresses one of the criticisms of library online catalogs—namely, that they are only useful for known items and do not facilitate discovery (Antelman, Lynema, and Pace, 2013).

Conducting research in classical music can be challenging for monolingual English speakers. Because so much of the classical canon was created by speakers of other Western European languages, one encounters foreign-language terms throughout the literature. Google Dictionary and Google Translate are two Google productions that can address this challenge. In order to quickly define a term, one need only type *define*: and the word of interest into Google's search box. Figure 5.1 shows the results of the search *define: fortissimo*. Beyond providing two definitions, Google Dictionary provides respective part of speech, and the pronunciation (in IPA and as a sound file) of *fortissimo*.

For terms that have not been assimilated into the English language, Google Translate can be a useful tool. Singers preparing a song in another language can type the text into the Google Translate box, allow Google Translate to detect the language, and select the desired translation language. A choir director preparing Brahms's *Requiem*, for example, could search the Brahms directive for the opening of the seventh movement, *Feierlich*, and find multiple translations, their part of speech, examples of usage, and a sound file for the pronunciation. See figure 5.2 for an example.

Specialized music dictionaries, bibliographies, and song/opera translation resources are well-researched and authoritative tools. Such resources are obviously preferable to the general dictionary, translation, and search tools offered by Google. Musicologist and librarian Jennifer Oates notes that "while a Google search might be helpful, published catalogues and bibliographies are far more comprehensive and often have already done the work for you!" (2014, 292–93). The challenge for the librarian, then, is to acknowledge the convenience of the Google tools, but to simultaneously encourage the user to go beyond these to the expert tools available in library collections.

Google Dictionary

For • tis • si • mo

/fôr'tisə,mō/

MUSIC

adverb & adjective
adverb: **fortissimo**; adjective: **fortissimo**
1.(especially as a direction) very loud or loudly.

noun
noun: **fortissimo**; plural noun: **fortissimi**; plural noun: **fortissimos**
1. a passage marked to be performed very loudly.

Origin

Italian, from Latin *fortissimus* 'very strong.'

Figure 5.1. Google *define* Search

MAPPING MUSIC

Google Maps can also assist in the translation and contextualization of music. Using the geographic tagging on Google Maps, it is possible to create thematic, publicly available maps that show the physical relationships between important music sites. These maps provide a free resource that can be created for a wide range of projects, from mapping special collections to engaging students in music history or popular music. Rap Genius (2014) has created Google Maps of locations mentioned in rap lyrics or with other rapper-related significance. The Rap Map in figure 5.3 is a Google Map of downtown Atlanta and marks locations mentioned by T. I., Jeezy, Ludacris, and other rappers.

In addition to mapping important music sites, Google Maps could also be used to help researchers identify collections of interest. The Irish Archives Map, for example, is a Google Map that lists archives throughout Ireland and provides some detail about their holdings. The open-access nature of these maps provides an opportunity for researchers to collaborate and share information about locations that would be of value to their particular research community. A music researcher could prepare a specialized map as a research plan and share it to solicit feedback from colleagues.

Google Translate

Translate

German - detected

Feierlich

English

Solemnly

Translations of **feierlich**

adverb	
solemnly	feierlich, ernst, gemessenen Schrittes, würdevoll, hoch und heilig, bei allem
adjective	
solemn	feierlich, ernst, festlich, ehrwürdig, erhaben, gravitätisch
grave	ernst, ernsthaft, ernstlich, gravierend, schwer, feierlich
ceremonial	feierlich, zeremoniell
festive	festlich, feierlich
formal	formal, formell, förmlich, offiziell, pedantisch, feierlich

Figure 5.2. Google Translate

GOOGLE+ FOR MUSIC RESEARCH

The social media platform Google+ allows for direct engagement between the library and its patrons. Libraries that share resources through social media create awareness of collections and also encourage a wider audience to use their collections. Because many libraries have already begun to participate in social media discussions, the inclusion of Google+ is a natural extension of those efforts and can be easily incorporated. For music collections, Google+ creates an opportunity to tie collections to discussions of particular genres, artists, composers, or other content. The community features allow for a single collection to exist in as many communities as are relevant to the held collections.

In addition to joining and sharing with communities, Google+ has integrated video features that include an archive for accounts. For music libraries, collections, and researchers with limited resources, the Google+ video archive allows for a web-accessible dynamic portal of recordings of music performances. This feature integrates with YouTube, the ubiquitous video-

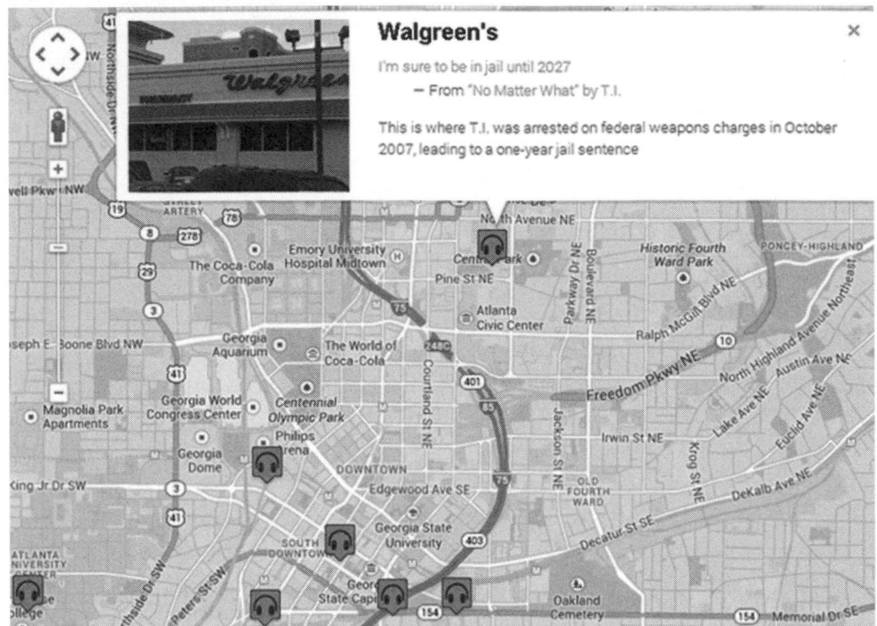

Figure 5.3. Google Maps—Mapping Music. ***Map data ©2012 Google, Sanborn***

sharing website already used by many libraries for the public sharing of video content.

As a social media platform, Google+ allows for direct engagement between the library and its patrons. Libraries that share resources through social media create awareness of collections and also encourage a wider audience to use their collections. Google Hangouts, a feature integrated into Google+ as well as other Google services, has been documented as an alternative to chat functions for librarians (Byrne 2012; Miller 2014). The opportunity to engage directly with the intended audience has also been embraced by musicians seeking wider exposure of their work (Schweitzer 2013). In a similar way, those interested in conducting music research can use the remote access to musicians to not only view performances but also to interview musicians. These interviews can then be disseminated or archived for future use by the researcher.

GOOGLE FOR MUSIC BUSINESS RESEARCH

A simple Google search for some music business topics will yield useful results from trade publications, newspapers, and magazines. However, the music industry does not typically disclose more information than is necessary. Some of the most useful statistics are compiled in the reports of the International Federation of the Phonographic Industry. These reports, including the *Recording Industry in Numbers*, are available for purchase at a cost that is likely prohibitive to most libraries. Individual companies can be researched using Google Finance, but the financial statements do not divulge more information than is available in the SEC filings. Google Public Data can be used to address the demographic and statistical information needs of those conducting market research in music business.

Historical data and secondary sources are much easier to come by. Google Books now includes magazines, including a complete run of *Billboard* from 1942 to 2011, in full view. This addition provides historical data on sound-recording sales, album and song charts by genre, market quotes, color advertisements, artist profiles, and more. Google Scholar can also be useful to music business researchers. As an interdisciplinary area of study, it can be approached from such disparate points of view as business, technology, copyright law, marketing, and music perception. As previously noted, Google Scholar yields relatively strong results with interdisciplinary topics. Unlike discipline-specific databases that must have comprehensive coverage in a single area, Google Scholar has great breadth, not depth.

CONCLUSION

Google Play Music, YouTube, Google Books, Google Scholar, and even simple Google searches facilitate music discovery, research, and performance inside and outside of our libraries. The investment in music metadata both through grants as well as private research by the company have made Google applications useful resources for musicians and music researchers. Librarians can use these powerful tools to meet users where they are and connect them with the rich resources in our collections. Google Books' link to Find in a Library and Google Scholar's inclusion of vendor link resolvers show Google's willingness to collaborate with libraries and to expose our resources to a broader audience. These features direct information traffic to libraries, but librarians are still learning how to integrate these powerful tools into the provision of information services. Gaining familiarity with Google's tools and products can only enhance the user experience and increase the exposure of library collections.

REFERENCES

Antelman, Kristen, Emily Lynema, and Andrew K. Pace. 2013. "Toward a Twenty-First Century Catalog." *Information Technology and Libraries* 25(3): 128–39.

Byrne, Richard. 2012. "Making the Most of Video in the Classroom." *School Library Journal* 58(8): 15.

Dediu, Horace H. 2013. "Measuring the iTunes Video Store." *Asymco*, June 19. www.asymco.com/2013/06/19/measuring-the-itunes-video-store/ .

Dougan, Kirstin. 2012. "Information Seeking Behaviors of Music Students." *Reference Services Review* 40(4): 558–73. doi:10.1108/00907321211277369.

———. 2014. "'YouTube Has Changed Everything'? Music Faculty, Librarians, and Their Use and Perception of YouTube." *College & Research Libraries* 75(4): 575–89. doi:10.5860/crl.75.4.575.

Garcia-Hernandez, Jose Juan, Claudia Feregrino-Uribe, and Rene Cumplido. 2013. "Collusion-Resistant Audio Fingerprinting System in the Modulated Complex Lapped Transform Domain." *PloS One* 8(6): e65985. doi:10.1371/journal.pone.0065985.

Giustini, Dean, and Maged N. Kamel Boulos. 2013. "Google Scholar Is Not Enough to Be Used Alone for Systematic Reviews." *Online Journal of Public Health Informatics* 5(2): 214. doi:10.5210/ojphi.v5i2.4623.

International Federation of the Phonographic Industry. 2014. "News." *IFPI*, March 18. www.ifpi.org/news/music-subscription-revenues-help-drive-growth-in-most-major-markets .

Johnson, M. 2012. "An Experiment in Music and Crowd-Sourcing." *Research @ Google*, May 4. http://googleresearch.blogspot.com/2012/05/experiment-in-music-and-crowd-sourcing.html .

Kemple, Steven. 2014. "Library Care Packages: CDs, Freshly Picked." *Library Journal*, April 9. http://reviews.libraryjournal.com/2014/04/media/music/library-care-packages-cds-freshly-picked-music-matters/ .

Miller, S. 2014. "Innovators: Virtually Connected." *Library Journal* 139(5): 50.

Oates, Jennifer. 2014. "Engaging with Research and Resources in Music Courses." *Journal of Music History Pedagogy* 4(2): 283–300.

Rap Genius. 2014. "Rap Map." Accessed June 10, 2014. http://rapgenius.com/map/ .

Schweitzer, Vivien. 2013. "Performing Together, Though Not Necessarily Onstage." *New York Times*, April 8, C4.

Scott, Rachel E. 2013. "The Edition-Literate Singer: Edition Selection as an Information Literacy Competency." *Music Reference Services Quarterly* 16(3): 131–40. doi:10.1080/10588167.2013.808941.

You, Shingchern, D., Wei-Hwa Chen, and Woei-Kae Chen. 2013. "Music Identification System Using MPEG-7 Audio Signature Descriptors." *Scientific World Journal* (January). doi:10.1155/2013/752464.

Chapter Six

Google Translate as a Research Tool

Andrew Wohrley

Google has created an enormous suite of tools for users, but after their original search engine, possibly the most important application is Google Translate (http://www.google.com/translate). Google Translate is available with Google Chrome and as an Android app that can be installed on Android devices from Google Play, and it is a powerful machine language translation system that was unimaginable even a decade ago. Unless otherwise indicated, this chapter will discuss the application as it exists on the Google Chrome browser or Android. There are other apps that work with other browsers, but for the sake of consistency, I will stay with the Chrome browser on the desktop environment.

The applications discussed in this chapter include:

1. Phrase translation online
2. Android app
3. Comparing news coverage of the same event in different languages, which is facilitated by Google News
4. Google Translator Toolkit
5. Google Website Translator Application in Espacenet
6. Translate document files

The simplest version of Google Translate is the word and phrase translation box. Google Translate works with regular HTML pages. For example, the citation:

Monique Desroches
"Les pratiques musicales, image de l'histoire, reflet d'un context."

> Un article publié dans l'ouvrage sous la direction de Jean-Luc Bonniol, Historial antillais. Tome I. Guadeloupe et Martinique. Des îles aux hommes, pp. 491–500. Pointe-à-Pitre : Dajani Éditions, 1981, 591 pp.

becomes

> Monique Desroches
>
> "The musical practices, image history, reflecting a context."
>
> An article published in the book edited by Jean-Luc Bonniol, Historial Caribbean. Volume I. Guadeloupe and Martinique. Island men, p. 491–500. Pointe-à-Pitre: Dajani Publishing, 1981, 591 pp.

For the monolingual, Google Translate offers point and click translations of over seventy languages.

Google Translate also works as an app on Android devices. Ideally, one can photograph foreign text and have it translated by Google. In practice the process of photographing the text with the app, and then finger swiping the text that needs to be translated is cumbersome, and not as satisfying as the results from online. If one does not like the translation one can take a second snapshot of the page and it is not uncommon for the automatic translation to be a bit better the second time. A comparison with *Paris Match* in online translation and in print might suffice to illustrate the difference. From online we get this from Google Translate:

> "Jean Dujardin Lovers Nathalie"
>
> Saturday, April 26, 20 hours at Boulogne, one hour before the show Nathalie. "That's what newspaper? "Says Jean the photographer, before adding:" Paris Match, class! "© DR
>
> April 29, 2014, Updated April 29, 2014
>
> By Pauline Delassus
>
> The new seductive French cinema does not hide his passion for the Olympic skater.
>
> Neither the ice nor the red carpet. . . . They chose the sidewalks of Paris to formalize their happiness. Five months after his separation from Alexandra Lamy, the star of "The Artist" has fallen under the spell of tap dance Nathalie Pechalat. As millions of viewers in the final free dance figure skating Olympics in Sochi. He endorsed the motto of his champion, "everything is possible, you have to believe and have no

doubt." This Feb. 16, the bubbly athlete had missed the medal but won a heart. After a few weeks of courtship, living in Japan at the World Championships, Nathalie began . . .

The Android translation out of the print *Paris Match* says:

spectace an hour before for any newspaper says Jean photographer. before adding a class Paris Match "Neither the ice nor the red carpet . . . They chose the streets of Paris to formalize their happiness!. Five months after his separation from Alexandra Lamy star of The Artist has fallen under the spell of tap dance Nathalie Pechalat. As millions of viewers of the final free dance art slip-Olympics in Sochi. He endorsed the motto of his champion, anything is possible, he must believe and not doubt be February the 16th, the bubbly athlete had missed the medal but won a heart. After a few weeks of courtship, living in Japan at the World Championships, Nathalie began

While these two translations are close, they do vary. The "sidewalks of Paris" become the "streets," "everything" becomes "anything," and most clumsily, "figure skating" becomes "art slip."

GOOGLE TRANSLATE USES

What are some uses of Google Translate beyond the obvious? One is that users can scan the web to see how different countries view the same event. Let's use the 2014 World Cup Final as an example and get our information from Google News. Some may not be aware that Google News produces national editions, with a website URL that follows the convention *news.google* and then the trailing letters are usually the two-letter international code for the nation (e.g., *fr*, *de*, *ar*, etc.). As can be seen in table 6.1, the different national perspectives are reflected in the national news coverage, and of course, some translations are smoother than others, and the Mexican translation in particular looks a little rough.

Seeking translations is also something that people want to do offline, too. The Android Google Translate app seems to work just as well as the online Google Translate. Here one sees Google at its best and at its worst. At its best, Google is a genius-driven corporate success story, constantly driving to improve its products and introduce new ones. At its worst, Google is a corporate behemoth doing what it wants and trusting that if its users are smart enough, they will catch up. Documentation, while effective (when it exists) for the technical questions it chooses to answer, leaves a lot to be

Table 6.1. Examples from Google Translate Covering the World Cup

Original	**Translation**
Germany Fußball-Weltmeisterschaft: Ex-Bundesminister Niebel verspottet Messi Handelsblatt-vor 8 Stunden Der deutsche WM-Erfolg wird auch von der Politik umjubelt. Für Misstöne sorgt Ex-Bundesminister Niebel mit einer spöttischen Bemerkung . . .	World Cup: Ex-Federal Minister Niebel mocked Messi Handelsblatt - 8 hours ago The German World Cup success is feted by politicians. For discord Ex-Federal Minister Niebel, makes a mocking remark . . .
Argentina Berlín celebra con los campeones del Mundial Deutsche Welle Español-hace 5 horas También será la primera copa que se quedará en Berlín. . . . espacio público del país para las transmisiones en vivo de los partidos del Mundial.	Berlin celebrates Champions World German wave Español - hace 5 horas It will also be the first drink that will stay in Berlin. Country . . . public space for live broadcasts of matches World.
Brazil O que deu certo e o que deu errado na Copa do Mundo 2014 no . . . Globo.com-15/07/2014 O Brasil voltou a receber uma Copa do Mundo após 64 anos e, apesar do pessimismo inicial – simbolizado pela frase "Imagina na Copa", que ...	What went right and what went wrong in the World Cup 2014 on . . . Globo.com - 15/07/2014 The Brazil returned to receive a World Cup after 64 years and, despite initial pessimism - symbolized by the phrase "Imagine Cup", which . . .
Spain Los jugadores alemanes se burlan de los argentinos en la celebración ABC.es-hace 1 hora Los germanos vencieron a la Albiceleste en la final de la Copa del Mundo y festejaron el triunfo ante cientos de miles de aficionados en Berlín.	German players tease Argentines in celebration ABC.es - 1 day ago The Germans beat the Albiceleste in the final of the World Cup and celebrated the victory before hundreds of thousands of fans in Berlin.
Mexico Peña Nieto confía que el Tri irá por la Copa del Mundo en Rusia 2018 CNN México.com-hace 4 horas No fue con la Copa del Mundo, como él se los pidió antes de que partieran a Brasil, sin embargo, agradeció su desempeño y les fijó una . . .	Peña Nieto Tri confident that go for the World Cup 2018 in Russia CNN México.com - hace 4 hours It was not the World Cup, as he ordered them before they left for Brazil, however, thanked their performance and set them . . .

desired. So some people do not know or appreciate that Google Translate will translate spoken words. To continue with the World Cup example, say "World Cup" into your Android phone and ask it to translate it into French and it instantly says, "Coupe de Monde," and spells the text out. For anyone who is in a foreign country and encounters a language barrier with a resident, as long as both parties have a little bit of patience, Google Translate can be very useful.

Also, if you are handy with spelling with your finger on an Android device you can spell words in a Translator-supported language and translate it to another language.

Google also offers Google Translator Toolkit, which requires a Google login and says, "Google Translator Toolkit allows human translators to work faster and more accurately, aided by technologies like Google Translate" (https://support.google.com/translate/toolkit/answer/147809?hl=en&ref_topic=22228). Clearly, Google wants people who require translation assistance, from the monolingual, to the professional translator, to use Google Translate. It is beyond the scope of this chapter to critique the usefulness of Google Translator Toolkit, but I have no doubt that it is an effective product, and in many cases, a professional translator, assisted by Google Translator Toolkit will be able to work faster with its assistance and provide smoother translations than just Google Translate.

The Google Website Translator allows website owners to install plug-ins to provide translations to other languages. While the Google Translate suite of products is undeniably useful, anyone who uses the Google Website Translator should be especially cautious to check that the translation does credit to the website and its owner, because clearly, some Google Translations are better than others.

There is a justifiable need for caution when dealing with the translations provided by Google. Reading the sports page, or using the Android App to order in a foreign restaurant will not tax a technical vocabulary, nor will the consequences be any worse than a misplaced order. But what if technical literature is being translated, and what if there is a legal issue involved? A good test of both issues is to use the European Patent Office's Espacenet database and test Google's translation of foreign patents.

Some background information is in order. The European Patent Office publishes patents in one of three official languages, German, French, or English. They also publish patents granted by select other patent offices, including the United States Patent and Trademark Office and the Chinese patent office. Many of these other patent offices publish in languages other than the EPO's three official languages, so automatic translations of patents are useful. The EPO says on its website:

> Patent Translate
>
> The EPO and Google have worked together to bring you a translation service optimised for patent documents.

Patent Translate is now complete and covers translations between English and 31 other languages, namely Albanian, Bulgarian, Chinese, Croatian, Czech, Danish, Dutch, Estonian, Finnish, French, German, Greek, Hungarian, Icelandic, Italian, Japanese, Korean, Latvian, Lithuanian, Macedonian, Norwegian, Polish, Portuguese, Romanian, Russian, Serbian, Slovak, Slovenian, Spanish, Swedish and Turkish. Translation from and into French and German is also available for 27 of these languages (www.epo.org/searching/free/patent-translate.html).

Espacenet may be searched at worldwide.espacenet.com, and a good example of something used around the world that would lend itself to patents is to search using the term *soccer ball*, which is a term used only in America and see what it finds internationally, which retrieves 1,115 results, mostly in the United States. Using the term football provides more international patents, almost all of which would be considered by Americans to be soccer.

An example of the Espacenet translate in action is in order. Here is a typical international patent:

Description: CN103816653 (A) — 2014-05-28

REBOUND PLATE AND REBOUND NET APPARATUS FOR BALL GAME TRAINING ENHANCEMENT AND HOME FITNESS

[The first four paragraphs read like this:]

[0001] 技术领域

[0002] 本发明涉及乒乓球、羽毛球、网球、足球、排球、棒球及壁球等的训练提高用回弹板和回弹网装置，同时可用于普通家庭室内与室外娱乐、健身。

[0003] 背景技术

[0004] 目前在体育运动如球类乒乓球，羽毛球，网球由于场地的局限，初学者无法快速提高掌握技术，难以 尽早建立对于该运动的兴趣与自信，目前训练界的产品设计主要是发球、捡球便利型，也无法在室内从运 动技术本身训练；普通家庭成员也无法作便利的健身运动，难以找到合适的运动伙伴，网络信息时代宅男 宅/女等越来越多，适于室内健身的器材呼之欲出；目前普通的个人用健身器材大多为力量型，如跑步机等 过于昂贵与笨重，市面上也没有轻便、灵活的运动器材可用于室内个人使用，无法满足人们的日常需要；灵动的才会更健康，也会更自信。

Figure 6.1. Typical International Patent

Translated by a click of the Translate button on the page, they read like this:

[0001]
Technical field
[0002]

This invention relates to table tennis, badminton, tennis, soccer, volleyball, baseball and squash training to improve network resilience with rebound boards and devices, and can be used indoors and outdoors ordinary family entertainment, fitness.

[0003]

BACKGROUND

[0004]

Currently in sports such as tennis balls, badminton, tennis, due to the limitations of the venue, beginners can not rapidly improve access to technology, it is difficult to establish as soon as possible for the movement's interest and confidence, currently training sector products designed primarily serve, picking convenient type, can not be in the room itself, from sports technology training; ordinary family members can not be conveniently exercise, it is difficult to find a suitable partner movement, network information age otaku / female, etc., more and more suitable for indoor fitness equipment ready to come ; the current ordinary personal use fitness equipment mostly for power type, such as treadmills too expensive and cumbersome, there are also no light-weight, flexible indoor sports equipment can be used for personal use and can not meet the daily needs of the people; facial will be more health, will be more confident.

While the translation is rough, the reader gets the sense that this is an apparatus that will allow people to train for sports individually. This example illustrates how useful Google Translate can be to individuals who may be expert in technology, and have a need to keep up with the patent literature but who do not have the time to pick up another language.

The example of Espacenet illustrates a larger point; there need not ever be a communications gap due to language. While there are still times and places where a professional translator should be hired to get the entire nuance of a document in a new language, right now Google Translate is "good enough" for people who can live with having a sense of the document.

Google used to have a plug-in for Firefox that incorporated Google Translate, but that has disappeared and leaves users with only Google Chrome for online support. Fortunately Google Chrome is an excellent browser and if you poke in the settings, under "Show advanced settings" you will see this message, "Offer to translate pages that aren't in a language you read." The checkbox by this line appears to be on by default, but if not, check the box and it will identify foreign languages other than the Chrome default and offer to translate them.

One final option for Google Translate is simply downloading a foreign language file to your own computer and then uploading it to Google Translate. If the file is PDF, this example will provide evidence that some formatting is lost and the translated text is garbled to an extent that would render it unfit for the purpose of translation. Here is what the header information for a Chinese language journal looks like:

不同脂肪源日粮对奶牛瘤胃代谢及
生产性能的影响
尹福泉1 嘎尔迪2 刘瑞芳2 于磊2
(广东海洋大学农学院1，湛江524088)
(内蒙古农业大学动物科学与医学学院2，呼和浩特010018)
摘要以3头年龄相同、平均体重为(650 4-20)kg、同处泌乳中后期平均泌乳量为15 kg左右、
装有永久
性瘤胃瘘管的荷斯坦奶牛为试验动物，试验采用3×3拉丁方设计，研究探讨向日粮中添加
不同的油料籽实
(日粮脂肪含量均在6．5%)对奶牛瘤胃代谢及生产性能的影响，为向日粮中添加油料籽实
提高乳脂中共轭亚
油酸含量的研究提供基础数据。3种籽实分别为葵花籽、亚麻籽、菜籽。研究结果表明，
日粮中添加不同的油
料籽实对奶牛瘤胃液纤毛虫数量动态变化、pH值动态变化、NH3一N浓度动态变化以及对
菌体蛋白浓度的影
响差异不显著(P>0．05)，对奶牛泌乳量及乳成分的影响差异不显著(P>0．05)，并均有改
善乳质的作用。

关键词畜牧学油料籽实瘤胃发酵饲料生产性能奶牛

Figure 6.2. Chinese Language Header Information

And here is what the translated header looks like:

May 2008
Volume 23 No. 3
Chinese Cereals and Oils Association
Journal
of the Chinese
Cereals
and Oils Association
V01.23, No. 3
May 2008
Different sources of dietary fat on rumen metabolism and
Impact on production performance
Yin Fuquan a Gaer Di 2 LIU Rui-fang 2
In

Lei 2

(Guangdong Ocean University of Agriculture and an Zhanjiang 524088)

(Inner Mongolia Agricultural University Animal Science and Medicine 2, Hohhot 010018)

Abstract three the same age, the average weight of (6504-20) kg, with an average lactation milk yield at around the late 15 kg, with a permanent

Holstein cows rumen fistula as experimental animals, the test using 3 × 3 Latin square design study to explore different oil seeds added to the diet

(Dietary fat content were 6.5%) on rumen metabolism and production performance for oil seeds add fat to the diet to improve the sub-conjugated

Research oleic provide the basic data. 3 seeds were real sunflower seeds, flaxseed, rapeseed. The results showed that dietary supplementation of different oils

Material changes in seed cow rumen fluid dynamics ciliates quantity, pH value of dynamic change, NH3-N concentration, and the dynamic changes of the protein concentration Yin shadow

Sound difference was not significant (P> 0.05), the impact of differences in milk yield and milk composition of dairy cows was not significant (P> 0.05), and have a role to improve milk quality.

Keywords animal husbandry oil seed production performance of dairy cows rumen fermentation feed

Here is a section of text:

Fat because of its high energetic value, as energizin[illegible]were sunflower seeds, flaxseed, rapeseed, were purchased from call
Low volume, after adding in the diet can increase energ[illegible]urbs. Experimental animals for three fitted with permanent rumen
Food intake, especially adding a certain percentage of [illegible] age, the averag[illegible]ight of (650
Fat, not only to ease the early lactation cows negative e[illegible]lactation.

Figure 6.3. Google Translate Flawed PDF Conversion

The garbling of the text is very disappointing because so much of Google Translate works simply and effectively, but on translating PDF files, it is prone to lose the page layout. Whether the fault is with Adobe or Google, this author cannot decide, but it is an issue that Google should correct.

Google Translate is a powerful tool for both the monolingual and the multilingual, but it is not perfect. For the monolingual it works best for giving a sense of the text that allows the reader to decide if he can use the text in his research and if necessary send it off for a more polished translation from a translator. For the multilingual, it has features that allow more automated assistance to their normal translation work. I think Google Translate may be the second most useful thing Google has done after search, and used wisely it can allow the researcher to consult international sources that they may never have considered before. In an increasingly globalized world informing oneself of what other people are saying in their own language is increasingly important. With Google Translate, one can monitor issues, technologies, and even what they are saying more easily than ever before.

Chapter Seven

Googling for Answers

Gray Literature Sources and Metrics in the Sciences and Engineering

Giovanna Badia

Google and its products complement bibliographic and full-text databases in the sciences and engineering by indexing gray literature sources; providing metrics for articles, authors, and journals; and allowing quick access to factual information. Over the past decade, first as a hospital librarian and currently as an academic librarian, I have employed various Google products for reference work, library staff training, and information literacy instruction. Through the presentation of real-life scenarios, this chapter will discuss

- techniques for using Google products and features to find facts, and gray literature sources such as patents, standards, and technical reports;
- creating a My Citations profile page in Google Scholar to calculate an author's h-index and track citations to his/her publications;
- building a Google Custom Search Engine to teach patrons and new library staff how to identify the best resource to search on a particular topic; and
- cases that illustrate how Google instruction can be included in information literacy sessions that are offered to different groups.

SOLVING DAILY PROBLEMS

A hospital employee once reported that the Windows desktop of one of the library's public access workstations was flipped upside down. I quickly typed the problem in Google's search box (i.e., *desktop turned upside down*) and found that pressing and holding down the Ctrl, Alt, and Up Arrow keys would correct the issue. This was not the first or last time that I used Google to solve computer and software problems, whether they were my own, the library's, or those of a patron. In addition to acting as my own personal computer technician, Google helps me with daily office tasks by performing the following roles:

- Calculator. Example: type *12*23* in Google's search box and press the Enter key to get the answer to this multiplication problem; use + for addition, - for subtraction, and / for division.
- Dictionary. Example: add *define:* before a word or phrase in Google to obtain possible definitions, such as *define: satisfice.*
- Event organizer. Example: I co-organize a brown-bag lecture series by and for librarians in my institution. Librarians usually discuss their research projects at these brown-bag lectures. I searched Google Scholar for *"McGill Library"* and clicked on the Create Alert link to be notified by e-mail whenever there is a publication or conference presentation by one of my colleagues that is indexed in Google Scholar. This information assists me in identifying librarians to invite as lecture speakers.
- Fact finder. Example: replace a word in a sentence with an asterisk (*) in Google to find web pages that contain the answer to a specific question, such as *there are * bones in the human body.*
- Teaching assistant. Example: type your search terms followed by *filetype:pptx* in Google to find PowerPoint presentations (created with Microsoft PowerPoint 2007 and onward) that contain your words, such as *searching for patents filetype:pptx*. This will list PowerPoint presentations on your topic, which may give you ideas for creating your own lecture or workshop notes.
- Unit and currency converter. Example: type *in* between the units or currency you wish to convert in the Google search box and press the Enter key to view the solution, like *100 kg in pounds* or *250 Canadian dollars in USD*
- Translator. Visit translate.google.com to translate words or entire documents from one language to another.
- Trip advisor. Example: use Google Maps for directions and to help choose hotels for a conference. I find inexpensive hotels in Expedia.ca for a specific city, and then check whether these hotels are close to the convention center by typing the address of a hotel and the convention center in Google Maps to get directions. On the ensuing map, I view what the hotel neighborhood looks like by selecting the satellite or street view option and then clicking on the hotel dot.

LOCATING AND SEARCHING FOR GRAY LITERATURE SOURCES

Google can also help with interlibrary loan and document delivery requests, especially when it concerns locating the full text of gray literature sources in the sciences and engineering. A frequently asked reference question is how to locate materials that the library does not own, such as conference proceedings. Students, faculty, and researchers will often obtain references to articles published in conference proceedings through searching the science and engineering databases that we subscribe to, for example, Compendex, Scopus, and Web of Science. These three databases index a large number of conference proceedings. Before asking the individual to submit an interlibrary loan request for the conference paper, I will search Google and Google Scholar to check whether a free copy is available online. Some associations post their conference proceedings on their website for free. I follow the same procedure when trying to locate a master's thesis or PhD dissertation since many universities have posted their students' theses and dissertations in institutional repositories, the majority of which are crawled by Google or Google Scholar.

Another frequently asked question that I receive is how to find the full text of standards, especially a specific ASTM standard. ASTM International (formerly known as the American Society for Testing and Materials) is an international organization that produces and publishes standards. Standards are procedures for making, measuring, and testing materials and products; they are created by different organizations around the world, including ASTM International, and can be as short as one page long. McGill University's Schulich Library of Science and Engineering owns the print version of the ASTM standards, which are issued on an annual basis and consist of a large multivolume set. Individual standards are grouped thematically and then by their codes in the multivolume set, which makes it challenging to find a standard by its code or title. Googling the code or title of the standard (e.g., *ASTM D4447-10* or *Standard Guide for Disposal of Laboratory Chemicals and Samples*) will point you to the record in ASTM International's online standards catalog, which lists the volume number in which the standard can be found.

Apart from locating the full text, Google can provide assistance in performing a search for gray literature sources on a topic, specifically when searching for technical reports and patents. Technical reports (i.e., research documents created by organizations in the public and private sectors) are not extensively indexed, if it all, by subject-specific or multidisciplinary databases, which makes them challenging to find. Googling a topic followed by *technical report* in quotes (e.g., *green airplanes "technical report"*) will find technical reports that organizations have made available on their websites. Searching for patents is a different story. Unlike technical reports, there are databases that provide comprehensive coverage of patents, such as Derwent Innovations Index. However, Derwent does not always provide the full text of the patent document or application for patents written in languages other than English. Searching Google Patents for the patent number will provide you with the English translation of the patent's description and claims if the patent is indexed by Google. Google Patents includes patents and applications from the United States Patent and Trademark Office, the European Patent Office, and the World Intellectual Property Organization ("About Google Patents" 2014). Google Patents supplements Derwent for finding the full text of patents and would be useful for patent searching on a topic when libraries do not subscribe to databases that index the patent literature. Additionally, patents are included in your search results, by default, when you perform a search on a topic in Google Scholar.

Since they index gray literature sources, Google and/or Google Scholar have been used to supplement database searches when performing a systematic review of the literature on a specific topic in the health sciences. For instance, a PubMed search for review articles published from 2009 to 2013 that contain the word *Google* in the article abstract obtained close to 1,400 results.

PROMOTING RESEARCHERS AND OURSELVES

Besides searching, librarians can employ Google Scholar to help promote researchers in their communities. Individuals can create a free My Citations user profile page in Google Scholar, where they can list their publications and make them public to the world. Type *Richard Feynman* in the Google Scholar search box and click on his name, under the heading *User profiles for Richard Feynman*, to view a sample user profile page. A My Citations profile page will display how many times each publication has been cited in Google Scholar, link to who cited each publication in Google Scholar, and calculate the total number of citations that the researcher's publications received, as well as calculate this individual's i10-index (i.e., the number of publications that have been cited at least 10 times) and h-index. The h-index is the number that corresponds to where the number of citations an author receives meets the number of the author's publications on a graph; for example, if Professor Smith's h-index is 20, this means twenty of his publications have been cited at least twenty times. To create or edit a My Citations profile page, click on the My Citations link at the top of Google Scholar's home page and log in to My Citations with your Google account username and password (or create a free Google account if you do not have one). There is a help link on the page to guide users in creating their own profiles. Librarians can make researchers aware of My Citations and help them create their own profile pages in order to make their work more easily findable online. Furthermore, librarians can utilize My Citations to acknowledge and draw attention to the scholarly contributions of library personnel in their institutions by creating a profile page for their libraries that would list the publications of all library staff members and then link to this page on their libraries' websites. Similar to this idea, the journal *Evidence Based Library and Information Practice* has created a profile page to show the most cited articles in the journal. Google *eblip*, click on the link for the journal, and then click on Most Cited Articles to view this profile page.

The article-level metrics—that is, the number of times cited and links to citing articles—which are displayed for each publication in a My Citations profile page, are also shown beneath each search result when you run a search on a topic in Google Scholar. There is additional information visible in Google Scholar for Web of Science subscribers. Thomson Reuters (the producer of Web of Science, a major collection of citation indexes) and Google Scholar formed a partnership in 2013 to cross-link their databases. Any member of an organization that subscribes to Web of Science and searches Google Scholar on his/her organization's authenticated network connection will also see, beneath each search result in Google Scholar, how many times an article was cited in Web of Science; clicking on this number will display the citing articles in Web of Science. To reciprocate, Web of Science provides a Look Up Full-Text link in each of its records; clicking on this link automatically runs an author and title search in Google Scholar to try to find the complete text of the publication. More databases may team up with Google Scholar in the future and, thus, add to the information that is currently available in the latter.

Along with finding article and author metrics in Google Scholar, you can browse a list of the most frequently cited journals in different languages and/or see the top English-language journals in a particular discipline by clicking on the Metrics link at the top of Google Scholar's home page. To view the top twenty journals in library science, for instance, you would click on the Social Sciences link on the left-hand side of the page and then click on Library & Information Science. Journals are listed by their h5-index. The h5-index is similar to an author's h-index but it applies to a journal over the past five years; for example, the h5-index for the *Journal of Academic Librarianship* is 27, which means that 27 articles published in this journal from 2008 to 2012 have been cited at least 27 times. Clicking on the h5-index number (27 in this case) for a specific journal will list the articles that were cited h times in the journal over the past five years (i.e., it will list the 27 articles that were cited at least 27 times in *The Journal of Academic Librarianship* from 2008 to 2012) and will show you the number of times each article was cited in Google Scholar and who cited it.

Google Scholar's metrics at the article, author, and journal level complement those found in traditional citation databases, such as Web of Science and Scopus, since they capture citations from gray literature sources that are not indexed in these databases.

FILTERING SEARCHES

Not all web pages indexed by Google are from authoritative sources. When librarians use Google to complement bibliographic and full-text databases, we filter the results for our patrons by selecting relevant and appropriate sources. Nevertheless, we are not available to answer reference questions 24/7, so how can we assist our communities in filtering Google search results?

One way is to create a Google Custom Search Engine, where you choose the web pages or sites that will be searched on a topic. Type *Google Custom Search Engine* in Google's search box, click on the relevant link, log in using your Google account username and password (or create a free Google account if you do not have one), and click on the Add button to start creating your own customized search engine. Help links are available on the page to guide you through the process. When you add entire sites, which contain databases, to your custom search engine, note that a Google Custom Search Engine will not retrieve as many search results as searching the original database on the website itself.

Even though your Google Custom Search Engine may not crawl all the pages on the sites you identify, I have found them useful in teaching university students and new library staff how to identify which resource to explore further on a topic. For example, I created a Google Custom Search Engine to help students in Earth and Planetary Sciences select which of several image databases were the best for finding images on specific topics, the results of which they could use in presentations and essays. My custom search engine (see http://bit.ly/EPSCse) included the image databases that I discussed with the students. Once they performed a search in the custom search engine and double-clicked on an image they liked, they would then be linked to the original website. I encouraged students at that point to perform a more comprehensive search on that site if they needed more images. I posted a link to the custom search engine, and the individual sites it searches, on the library's subject guide for Earth and Planetary Sciences.

Similarly, I had previously created a Google Custom Search Engine to help train new library staff in the hospital by grouping together authoritative sites that were useful for answering frequently asked ready-reference questions (i.e., factual questions that usually take less than ten minutes to answer) that they received at the library services desk, such as questions about finding basic consumer health information for a disease, finding guides to citation styles, finding statistics, and so on. The purpose of the custom search engine was to help new library staff quickly and confidently identify the best vetted site to search for answering a specific reference question.

TEACHING GOOGLE

If librarians utilize Google and its products to find facts, gray literature, and/or citation information, they can also teach others how to do this, as well as how to search Google effectively and read its search results critically. Google and/or any of its products can be the topic of a stand-alone information literacy session, or it can be incorporated into a workshop or lecture on a broader topic. Including effective Googling in an information literacy session will draw participants' attention since many people use Google and/or one of its products on a regular basis.

My first Google, stand-alone, information literacy session was a fifty-minute lunchtime lecture that was offered to hospital employees at Montreal's Royal Victoria Hospital in April 2006. The lecture was titled "Googling the Best Health Care Information," and it covered advanced Google search techniques for bringing relevant health care information to the top of Google's search results, Google Scholar, Google Images, and Google Alerts, as well as discussed the advantages and limitations of using Google over the PubMed/MEDLINE database. I presented different information scenarios in the lecture, and my overall message was that it was definitely appropriate for audience members to use Google in certain cases, and databases like PubMed/MEDLINE in other cases, for meeting their information needs; both had their place in the health care environment. The lecture was enthusiastically received by a large and mixed audience; approximately one hundred fifty individuals attended, consisting of different health care professionals, hospital administrative and support staff, and researchers.

Its overwhelming success was partly due to the fact that the timing for a Google lecture was right, since there had recently been published articles in the medical literature (which generated buzz) about how Google was affecting medicine (e.g., Giustini 2005; Henderson 2005; Turner and Purushotham 2004). I was asked to repeat the lecture by different groups in the hospital to accommodate those who had not been able to attend, and I built upon it with a new advanced lecture on Google Scholar in February 2008. The latter was also well received and asked to be repeated. The key to creating attractive lectures on Google, its products, or any other resource is to include relevant content that directly addresses individuals' daily tasks, information challenges, and/or interests.

In my current position at McGill University, I incorporate Google and/or its products in my information literacy sessions whenever possible so that I can engage and connect with my audience. I frequently teach undergraduate engineering students how to search for peer-reviewed journal articles on their topics in ninety-minute, hands-on workshops. One of my workshop activities requires students to brainstorm synonyms for the concepts in their research topics. I tell students that they can Google their concepts to find synonyms or different ways of expressing each one, and I discuss *Wikipedia* as a starting point for obtaining a basic understanding of their research topics that may lead them to refine their topics and database search strategies. After teaching students how to create effective search strategies to increase the number of relevant results they obtain, and how to run their strategies in databases that cover the engineering literature such as Compendex, I show them how they can apply their strategies to searching Google Scholar. I will also teach students how to export their search results from Google Scholar to EndNote if an EndNote component is included in the workshop.

Acknowledging that your audience will use Google and/or its products in their research means that you can demonstrate how Google and bibliographic databases complement each other, thereby increasing the relevancy of your information literacy session for your audience.

GOOGLING IN THE SCIENCES AND ENGINEERING

Google and its products can be utilized effectively in the sciences and engineering to complement bibliographic and full-text databases, specifically when it involves finding factual information and gray literature sources, identifying highly cited publications, promoting researchers through the creation of online user profiles, and teaching others how to search the published literature.

REFERENCES

"About Google Patents." 2014. *Google*. https://support.google.com/faqs/answer/2539193?hl=en .

Giustini, D. "How Google Is Changing Medicine." *BMJ (Clinical Research Ed.)* 331, no. 7531 (2005): 1487–88. doi: 10.1136/bmj.331.7531.1487.

Henderson, J. "Google Scholar: A Source for Clinicians?" *CMAJ: Canadian Medical Association Journal* 172, no. 12 (2005): 1549–50. doi: 10.1503/cmaj.050404.

Turner, M. J., and A. D. Purushotham. "Accidental Epipen Injection into a Digit—the Value of a Google Search." *Annals of the Royal College of Surgeons of England* 86, no. 3 (2004): 218–19. doi: 10.1308/003588404323043391.

Chapter Eight

Legal Research Using Google Scholar

Ashley Krenelka Chase

Legal research can be both complicated and expensive. Google Scholar is one of the best free resources for performing research of U.S. case law, law review articles, and patents. When Google published a blog post announcing Google Scholar's inclusion of case law in 2009, Google indicated that the database would include both well-known legal cases such as *Plessy v. Ferguson* and less-common cases that affect people throughout the United States (Acharya 2009). In addition to the expansive inclusion of cases from all jurisdictions, the legal content on Google Scholar is updated consistently on a weekly basis, making it a trustworthy source. Google Scholar is both comprehensive and user friendly; the search engine looks and feels like a normal Google search. Librarians and patrons in all libraries will benefit from having Google Scholar at their disposal to perform basic legal research.

Researching case law, law review articles, and patents in Google Scholar requires essentially the same process. Each search can be tailored differently, however, depending on the information sought. As with a typical Google search, legal research can be done from a primary search screen or an advanced search screen. The search can then be narrowed by user preference, depending on the search performed. In addition, citation trackers and alerts can be set up for future reference.

RESEARCHING U.S. CASE LAW

Case law is primary legal authority. Because the United States has a common-law legal system, courts must follow decisions from previous cases on the same legal topic. Performing case law research is essential to knowing what the law is in a given area. Google Scholar allows users to research federal and state case law in one place. Legal research is most easily accomplished in two ways: keyword searching and searching by party name. Researching U.S. case law by either method is really quite simple, as is narrowing the results when the search is complete. By simply selecting Case Law in the primary Google Scholar search screen, you are ready to begin.

Searching

Keyword searching. Keyword searching for legal cases in Google Scholar works identically to searching for anything else in Google. Performing legal research with this search engine requires no special knowledge of the law, or advanced research techniques. The user enters keywords relevant to the search, then clicks the magnifying glass to see the results. When performing legal research in any search engine, users should try to search for language related to the kind of case they need. For instance, if you are performing a search for cases where the court addressed Section 61.08 of the Florida Statutes (which deals with alimony), it would be beneficial to run a Google Scholar search for *61.08 alimony*. This simple search, with no other details and without narrowing, yields over 850 relevant cases.

It is unnecessary, however, to know related statutes or technical terms when performing case law research. A simple search for murder yields well over 500,000 cases. A further-defined search for first-degree murder produces over 270,000 results. By adding the words *gang* and *Ohio* to this keyword search (for a search that reads: *first-degree murder gang Ohio*) reduces the results to just over 1,850. By adding or subtracting words that are relevant to the search, Google Scholar users can find relevant case law from state and federal courts.

Party name searching. Often you will hear about a case by party name, such as *Roe v. Wade*, *Bush v. Gore*, or *Marbury v. Madison*. While famous court decisions such as these can be found all over the Internet, less well-known decisions may be hard to find with an average search engine. Google Scholar allows for easy searching by case name. There are two ways to accomplish party name searching. The first is to enter the names of the parties into the search engine and to let Google determine which case best fits your search. For example, if you are looking for the United States Supreme Court case *Miller v. Alabama*, searching for either *Miller Alabama* or *Alabama Miller* will yield the same result. The U.S. Supreme Court opinion is the first result in the list either way. Searching for party names in this fashion is not much different than a keyword search.

If, however, you are searching for a case that is less well-known (as most cases not heard by the Supreme Court are), adding the *v* to the search may be helpful. Adding the *v* in between the party names allows the search engine to recognize that you are looking for cases with those parties, which produces better results. As an example, if you are looking for *Heckman v. State*, but search simply for *Heckman State*, Google Scholar lists over 290 results, none of which are the case you want. If, however, you run a search for *Heckman v. State*, the search engine returns the same number of results, but prioritizes the cases with those party names on the first page. This allows the user to find results faster, without a significant change in search strategy. In order to tailor this search further, it is important to narrow the results appropriately, and sort them based on needs.

Narrowing and Sorting

Jurisdiction. Narrowing by jurisdiction is one of the most important tools in any legal researcher's toolbox. Because of the nature of the law, the jurisdiction in which a case is decided is binding on all future cases in that jurisdiction. Determining the law in your jurisdiction, and in any other jurisdictions that have an impact thereon (such as the United States Supreme Court), is essential to performing legal research. If we continue to use *Heckman v. State* as an example, our Google Scholar search shows cases with those party names in several Florida jurisdictions. The tool bar on the left of the page allows the user to narrow to only the jurisdictions that are important to this search. Clicking Select Courts on the left of the page brings up a list of every federal and state court in the United States. From here, the user can select the jurisdiction that is most important to him or her. By selecting Florida Courts, the search is narrowed from over 290 to 4, a much more reasonable number.

Of course, narrowing by jurisdiction can also be done in a keyword search. A search for *invasion of privacy* in Google Scholar yields over 7,000 results. Using the Select Courts feature and choosing the D.C. Circuit (under Federal Courts), the search can be narrowed to cases only within that federal jurisdiction, bringing that number to just over 1,500 results. That many results is still a lot, for first-time legal researchers and expert legal researchers alike. Additional narrowing by date can be performed to further tailor the legal information retrieved.

Date. Determining an appropriate date range is essential for performing most research. Legal research is no different. Often, reviewing and analyzing the most recent cases on an issue can provide excellent context for the state of the law in a particular area. Reviewing current cases can also provide insight into the history of the law in that area, since cases are decided based on previous decisions within that jurisdiction. Narrowing the jurisdiction first, and then the date in Google Scholar can yield excellent, highly relevant results.

In the example above, a Google Scholar search for *invasion of privacy* in the D.C. Circuit yields over 1,500 results. When those cases are narrowed to cases "Since 2014" on the left hand toolbar, the user is left with only 8 results! This is an ideal situation for any user performing legal research for the first time. As another example, a case law search for *slip and fall* leads to just over 169,000 results. If the user is looking for a specific slip-and-fall case she knows was decided in 2011, she can narrow by selecting Custom Range in the tool bar on the left and inserting 2011–2011. The results are further limited to over 3,000.

At this point, the user may want to re-sort the results. The left tool bar offers the option of sorting by relevance or by date. If the researcher knows the case she was looking for occurred toward the end of 2011, searching by date may be helpful, as the cases are ordered chronologically, beginning with the most recent. While the cases are reordered, there are no fewer of them. The user may want to narrow the jurisdiction at this point, or add some additional, advanced search terms to narrow the results list even further.

Advanced Searching

In the slip-and-fall example above, the search narrowed to 2011 resulted in far too many cases for a reasonable researcher to review. Even the most advanced legal researcher cannot sort through over 3,000 cases to find the one he is looking for. Advanced searching for case law in Google Scholar is simple, and can be done from the results page after other sorting, by either date or jurisdiction (or both!), has been completed.

The search bar, on the top of the Google Scholar search page, contains a small, downward-pointing arrow on the right. Selecting this arrow opens an advanced search box. The advanced search box allows users to enter all of the words they are looking for, including exact phrases. The advanced search box also allows the user to indicate words she does not want to see in her results. A word of warning: novice legal researchers should rarely enter words into the Without the Words search box, as courts tend to use words in different ways than the average person is accustomed to. By eliminating words from the search right away, the user runs the risk of eliminating exactly what he needs!

To perform an advanced slip-and-fall search in Google Scholar, select the downward-facing arrow. Enter the words *slip and fall* in the With All of the Words box. Enter the words *social security* in the With the Exact Phrase box. Enter the word *disability* in the With at Least One of the Words box. Make sure that you enter 2011 in both boxes for the Return Articles Dated Between boxes. Without narrowing the jurisdiction, the advanced search brought the results down from over 3,000 to 86. If, from there, the user narrows to the jurisdiction of interest to her, perhaps the Ninth Circuit, her search can be narrowed from 86 to 8. The results become similarly narrow for every jurisdiction. Advanced searching in Google Scholar is a quick, easy way to tailor your results from the beginning, to avoid an overwhelming number of cases turning up in your results.

Citation Tracking and Alerts

Both citation tracking and alerts are particularly useful in performing legal research. If a legal researcher relies on a specific case to make an argument, and then that case is overturned, the argument is no longer relevant and the researcher looks silly. A smart researcher looks at how cases are cited and implements citation tracking to be sure that he is only using good law. In addition, an adept user may want to set alerts to be notified of new information that becomes available resulting from certain search terms.

Citation tracking. Citation tracking is the single most important activity in performing legal research. If a user finds a case that is directly contradicted, or even overturned, by another case in the same jurisdiction, the case he wants to use is either no longer good or needs to be discussed in a context with the related cases. Performing a search for *Perry v. Schwarzenegger*, a Northern District of California Case, in Google Scholar will pull up the relevant case. After the user clicks into this case, a How Cited button appears under the case citation at the top of the page. When the user selects How Cited, a page appears that gives relevant quotes about how *Perry v. Schwarzenegger* has been discussed in other cases. On the right, boxes appear that say Cited By and Related Documents. The Cited By list contains all cases in every jurisdiction that have cited to *Perry*. The Related Documents list contains documents, not all of them cases, that reference or relate to *Perry* in some way.

It is important to note that the cases listed in the Cited By box do not say whether the new case overturned or disagreed with the previous case. This puts the onus on the user to read each opinion (in the jurisdiction for which they are doing the research) to determine how the cases that cite *Perry* have been influenced or use the original opinion. The ability to perform this level of research on Google Scholar is unique among all free legal search engines. Most do not offer this advanced feature!

Alerts. Often, legal researchers will run the same search over and over again in an attempt to find the latest and greatest case on their issue. Google Scholar has the option of creating an alert for any search, cutting down on the amount of time needed to run and rerun searchers. To set an alert, the user should run the case law search (or advanced search) as she normally would in Google Scholar. In the tool bar on the left, the user can select Create Alert (which appears with an envelope next to it). The user must simply add her e-mail address to the e-mail field and indicate how many results she would like to see in the e-mail. From there, Google Scholar will alert the user to any and all new information related to his search!

RESEARCHING LAW REVIEW ARTICLES

Law review articles are secondary legal sources, often used to find primary authority (such as case law) on a particular topic. If a user is unfamiliar with a particular area of the law, secondary sources can be helpful in giving an overview of the topic to be covered. Google Scholar allows users to research law review articles in the same manner as keyword searching for U.S. case law, and many of the above research tips apply to searching for law review articles as well. From the Google Scholar main page, a user needs only to select Articles (and uncheck the box for Patents) to begin looking for law review articles.

Keyword searching for law review articles in Google Scholar works identically to searching for anything else in Google or Google Scholar, including U.S. case law discussed previously. Searching for law review articles with this search engine does not require any special knowledge of the law or advanced research techniques. The user enters keywords relevant to the search, then clicks the magnifying glass to see the results. Again, when performing research of this kind, users should try to search for language related to the kind of law review article they are looking for. For instance, if you are performing a search for articles that have to do with adoption of children, it would be beneficial to run a Google Scholar search for *law adoption of children Russian*. Including the word *law* in the research string increases the probability of law review articles being included as the most relevant results.

The same advanced searching and narrowing techniques that applied to researching U.S. case law also apply to researching law review articles. Using the downward arrow in the search box, the user can add or subtract any legal terms or phrases she wishes. She may also specify the time frame from which she would like the articles to have been published. Again, this sort of tailoring makes Google Scholar an ideal search engine for those without access to expensive legal research databases. Alerts can also be set for law review articles in the same manner as they are set for U.S. case law.

RESEARCHING U.S. PATENTS

Google offers two platforms for researching U.S. Patents: Google Patents and Google Scholar. While Google Patents is a slightly stronger database for research of this kind because it searches for only patents, Google Scholar may also be used. To search for a patent on Google Scholar, check the box for Include Patents when performing the search. It is also important to include the word *patent* in any search terms input into Google Scholar. This ensures that patent results will be identified as the most relevant. If the user is looking for a patent created by a specific inventor, it is also useful to include that inventor's name, in quotations, in the search terms.

As an example, to find the patent specifically created by Erik Reader, the user may want to enter the search terms *patent "Erik Reader"* into Google Scholar. This search yields one result, and it is the patent for a "foot held waste basket" invented by Anthony C. Baier and Erik W. Reader, published in 2007. Adding the quotation marks around the inventor's name ensured that any result would have that name in it. In addition to these tricks, the same advanced searching and narrowing techniques detailed above can be applied to patent searches. The user should always remain mindful of adding the word *patent* to the search, regardless of whether the search is standard or advanced. Alerts can also be added for patent research in the same manner as they are set for U.S. case law and law review articles.

SUMMARY

Google Scholar is an extremely useful tool for performing legal research. It is particularly useful in libraries where patrons ask for legal information but traditional legal databases are unavailable. Because Google Scholar contains the same U.S. case law, patents, and law review articles that can be researched using traditional, paid resources, it is an extremely useful tool for libraries that cannot afford another expensive database. Google Scholar allows users to perform basic or advanced searches, to track citations, and to set up alerts so they can remain informed about the legal issues they research. The intuitive interface with which Google users are already familiar is unintimidating, and users are likely to feel at home performing legal research like any other method of research they have performed on Google in the past.

REFERENCES

Acharya, Anurag. 2009. "Finding the Law That Governs Us." *Google Official Blog* (blog), November 17. http://googleblog.blogspot.com/2009/11/finding-laws-that-govern-us.html .

Part II

User Applications

Chapter Nine

Better Images, Better Searchers

Google Images and Visual Literacy in the Sciences and Social Sciences

Melanie Maksin and Kayleigh Bohémier

This chapter describes our instruction efforts concerning Google Images, a specialized image search engine. We were inspired to teach Google Images to an academic audience by our experiences in the Power Searching with Google MOOC, Google's effort to improve searchers' understanding of their platform's capabilities, and by our academic community's interest in finding images for coursework, presentations, publications, and other scholarly activities. Our workshop evolved from focusing on a single search tool into an exploration of diverse electronic resources for visual materials and strategies for effective image searching, with Google Images at the center of a richer visual literacy conversation. Treating Google Images and other Google products as a part of a balanced instruction session leads attendees to more reflective query creation, an improved understanding of Google's context in an academic setting, and a better command of digital image resources available at our institution.

THE FIRST GOOGLE IMAGES SESSIONS

During fall 2012, we began offering half-hour workshops focused on academic uses of Google. Following the structure of the Power Searching with Google MOOC, we separated Google search platforms (Scholar, News, Images, Books) into different sessions but intentionally emphasized the search operators and techniques common to all Google interfaces. The Google Images session highlighted the platform's unique features, as well as an overview of basic and advanced search operators, suggestions for constructing meaningful searches, and the caveat that Google is just one tool of many in a researcher's arsenal. Although one Google Scholar workshop was aimed at undergraduates, most sessions, including Google Images, were intended for a broad audience of undergraduate and graduate students, faculty, staff, and other researchers around campus.

Advertising these first Google workshops proved challenging. Attendance was low—or, in the case of our Google Images session, nonexistent. We promoted the series as "Beyond the Basics," but we suspected that recommending instruction in such familiar tools might feel patronizing to those who assume that they already know how to use Google effectively. We also wondered if, by attempting to attract "everybody," we were missing an opportunity to craft a targeted, relevant session for a narrower audience. The following semester, as we grappled with ways to reframe our Google workshop series, participants in our second attempt at a Google Images session provided very valuable feedback. One participant, a science graduate student seeking images to incorporate into presentations, commented that he had not expected to learn anything new from a workshop on Google Images, yet had been pleasantly surprised. Participants were also curious about digital images from local special collections. For our audience, the appeal of this workshop was not its focus on the Google Images platform, but the opportunity to learn about tools and strategies for locating and making use of images for specific academic purposes. We decided to assess what we taught in the workshop to improve its value to attendees.

In fall 2013, we reintroduced the workshop as "Finding Images for Scientists and Social Scientists." This broke with our "Google [Scholar, Images, News]: Beyond the Basics" naming convention and promised participants a more holistic approach to image searching. We increased the session from thirty minutes to an hour and added content about finding images using subscription databases and digital collections from libraries, archives, and other cultural institutions. Google Images remained at the core of the workshop because potential attendees are familiar with the Google search system, and teaching them how to navigate Google Images effectively can translate to better search strategies in other Google products and even in non-Google image databases.

While broadening the scope of search tools and techniques addressed in the session, we narrowed our marketing to researchers in the sciences and social sciences. As librarians at Yale University's Center for Science and Social Science Information, we support researchers in these disciplines, and they frequently require images for use in publications, presentations, and teaching and learning in the classroom. An environmental scientist studying climate-game interactions in the pre-Columbian Americas might seek a photograph of bison. An astronomer may want an image of the H-R diagram, which describes stellar type and evolution, for use in an undergraduate class. A sociologist studying riots in the Middle East may need contemporary images to include in a conference presentation.

Many students and scholars in the sciences and social sciences are familiar with creating or interpreting visual elements in their research, typically in the form of tables, maps, graphs, or other data visualizations. However, their daily scholarly practice does not often provide them with established methods for learning about finding and using images. We intended for this session to expand participants' understanding of the image resources available and to confront questions that can arise when searching for visual images to incorporate into academic work. Although a "Finding Images for Humanists" session would encompass similar concerns, students in these disciplines are more likely to encounter images in the classroom or through course-integrated library instruction, and advanced scholars will already have an intimate familiarity with relevant image resources (e.g., ARTstor) and an understanding of the disciplinary norms for describing and contextualizing art, cultural artifacts, and visual media. By targeting researchers in the sciences and social sciences, we addressed their academic context by focusing on the tools and techniques that would have the greatest significance for them.

GOOGLE IMAGES AS A SPRINGBOARD FOR VISUAL LITERACY

The redesigned "Finding Images" session featured some of the instructional content from our first Google Images workshops. We began the session with Google Images and asked participants: "How would you search for images related to women in science, technology, engineering, and mathematics?" This started a discussion about creating keyword searches and using operators effectively. Most attendees were familiar with phrase searching (using quotation marks, e.g., *"women in stem fields"*), but few had tried synonym searching (using a tilde, e.g., *~women science*) or excluding a term for the purpose of disambiguation (using a minus sign, e.g., *women stem -cells*).

With the results of the sample search, we highlighted important features in Google Images such as filtering by color, image type, file type, or size and indicated how these filters can dramatically cut down on unwanted information noise in a search. We recommended limiting a search to a specific domain (e.g., *site:edu* or *site:nature.com*) to retrieve a more focused set of image results.

We also showed attendees how to search by image, a newer Google Images feature. Similar to the reverse image search engine TinEye (www.tineye.com), but with the benefit of the familiar Google interface, the camera icon in the Google Images search box prompts the searcher to type in an image URL or upload an image to find a match. While helpful in academic contexts for identifying images of unknown provenance, we pointed out that this could also be helpful for personal information needs. For example, a popular image circulating during Hurricane Sandy showed the Statue of Liberty in the foreground of an impressive cloud structure. Uploading a copy of this image can help an individual discover that the doctored photo originated several years prior to the hurricane (Mikkelson and Mikkelson 2012).

When we moved from Google Images to other image resources, we provided a bridge between Google Images–specific strategies and those that also apply to a plethora of free and subscription databases. We chose a manageable selection of image resources:

- Yale University Library's Digital Collections (http://digitalcollections.library.yale.edu/) for its local appeal;
- the New York Public Library's Digital Collections (http://digitalcollections.nypl.org/) for another resource of great depth and breadth; and
- Springer Images (www.springerimages.com/) for its relevance to researchers in the sciences and as an example of a subscription database.

In each resource, we recommended techniques for effective searching. Some overlapped with techniques in Google Images, but others revealed contrasts—typically related to subject terms' utility and the presence of quality metadata in the other platforms. We also proposed strategies for determining an image's source and locating statements related to copyright. The session wrapped up with further comments on copyright and resources for determining how an image can be used (e.g., Wikimedia Commons and the Copyright Clearance Center), and we returned to Google Images to demonstrate using the site: operator to narrow a search to repositories of public domain and Creative Commons images (e.g., *site:commons.wikimedia.org "women in science"*).

In recasting the Google Images session as "Finding Images for Scientists and Social Scientists" and focusing less on the elements specific to a single search engine, we found that the *ACRL Visual Literacy Standards* could inspire a number of approaches to image discovery. In response to the growing emphasis on visual media in scholarly practice and trends in how students discover and use (or misuse) visual materials, the *Association of College and Research Libraries (ACRL) Visual Literacy Standards for Higher Education* were approved and disseminated in 2011 as a complement to the *Information Literacy Competency Standards for Higher Education*. While the *Information Literacy Standards* promote the enrichment of students' abilities to find, contextualize, evaluate, and use information, these standards do not articulate the specific issues posed by visual materials. Images and visual media can have many possible functions and interpretive frameworks; these materials are both information and "aesthetic and creative objects" (Hattwig et al. 2011, para. 7). Visual materials carry unique challenges in terms of ethical and legal use, particularly related to reproducing, sharing, and modifying or remixing—any of which might be of interest in academia, but all of which require special caution. The *Visual Literacy Standards* propose a "structure to facilitate the development of skills and competencies required for students to engage with images in an academic environment, and critically use and produce visual media throughout their professional lives" (para. 6).

Hattwig, Bussert, Medaille, and Burgess (2013) provide a thorough overview of the current visual literacy landscape after the introduction of the *Visual Literacy Standards* and suggest approaches to visual literacy instruction. They encourage librarians to look for new roles related to the standards, specifically related to the use of visual media in academic work (e.g., through collaboration with faculty and information technologists), but note that "[finding] and accessing images may be the aspect of visual literacy most familiar to the greatest range of academic libraries and librarians" (Hattwig et al. 2013, 77). Of the seven standards established in the *ACRL Visual Literacy Standards* document, there are five that resonate with the goals of our redesigned workshop.

In table 9.1, we map the content of the "Finding Images" session to the *Visual Literacy Standards*. From the table, it is clear that Google Images is still a valuable tool in our "Finding Images" workshop. So much of the session's instructional content takes Google Images as a starting point—how to structure a search query, how to employ search limits or filters to refine a search, the importance of provenance and adherence to copyright—that it is difficult to envision a "Finding Images" session without it. However, with the addition of new resources and the use of Google Images for context and comparison, the "Finding Images" workshop is more closely aligned with visual literacy outcomes.

Although one workshop cannot produce a fully visually literate researcher, the questions and strategies employed in this session can serve as either an introduction to or a reinforcement of skills for effective image searching. This session could synergize with classroom instruction, independent or group research experiences, or other academic or personal activities that involve visual media.

WHY TEACH GOOGLE IMAGES?

When presenting Google Images as a resource for academic research, we considered broader pedagogical goals beyond the *Visual Literacy Standards*. We hoped to inspire participants to think critically about search tools and

Table 9.1. Visual Literacy Standards for "Finding Images for Scientists and Social Scientists" Instruction

Question Posed	Resource Used	Strategy
Standard One: "The visually literate student determines the nature and extent of the visual materials needed."		
What terms will you use to search for the image?	Google Images	Brainstorm keywords
What kind of image do you need?	Google Images	Filter by image type (photo, line drawing, etc.), file type (filetype:jpg), size, color
	Springer Images	From the advanced search interface, select image type or color
Are digital images adequate, or will you need to explore an archive to find relevant visual materials?	Yale University Library's Digital Collections and the New York Public Library's Digital Collections	Look for information about collections in digital image databases; these digital collections are just a sliver of what's available to on-site researchers
Standard Two: "The visually literate student finds and accesses needed images and visual media effectively and efficiently."		
How can you adapt your search terms/ strategy to suit a particular image database?	Google Images	Use operators like OR to expand a search and ~ to find synonyms; employ additional operators or filters
	Yale University Library's Digital Collections, NYPL Digital Collections	Begin with a keyword search then explore further using assigned subject terms
	Springer Images	Browse by subject; search by keyword in caption or description; limit by publication source, publication date, etc.
Is there a relevant subject- or discipline-specific image database?	Yale University Library Visual Resources Guide (http://guides.library.yale.edu/images)	Consult Yale's guide to visual resource databases or contact a subject librarian for recommendations
	Springer Images	Compare images from this collection to items found via Google Images
Standard Three: "The visually literate student interprets and analyzes the meanings of images and visual media."		

Are there clues within the database that can help you learn more about the image?	Yale University Library's Digital Collections, NYPL Digital Collections	View image details to determine which library collection or archive houses the original image; subject terms, names, and other metadata provide additional context
Standard Four: "The visually literate student evaluates images and their sources."		
What is the image's provenance—where did it come from and where else is it used?	Google Images	Reverse image search (image from Pinterest without attribution; drag and drop a JPG file)
	Springer Images	View details about the image's publication source and view the image in the context of the article/book
Is the image from a reputable source?	Google Images	Use a site: search to find image content from trusted domains and sites
Standard Seven: "The visually literate student understands many of the ethical, legal, social, and economic issues surrounding the creation and use of images and visual media, and accesses and uses visual materials ethically."		
What are the copyright or fair use implications for this image?	Google Images	Search by usage rights
	Yale University Library's Digital Collections, NYPL Digital Collections, Springer Images	Locate statements related to copyright and citation formatting in image databases
	Copyright Clearance Center	Resources for understanding copyright and seeking permission to use published material
Can public domain or Creative Commons sources meet your image need?	Google Images	Search by usage rights
	Wikimedia Commons	View an Image's "File" page for source and license information

processes, and to make connections between their experiences with Google Images and the wider information landscape. In particular, we sought to encourage mindful searching and to build a bridge between Google and the library's many databases.

As we pass through the 2010s, we see the prevalence of mobile technology and search; an overabundance of content due to the ease with which it is created; and the emergence of reflexive searching in individuals' daily lives. "Reflexive searching" refers to the instinct many have to reach for a smartphone or computer keyboard to execute a quick Google search. Such searches are either unstructured, stream-of-consciousness jumbles or very short queries. As Choi's (2010) study of undergraduate image searchers bears out, brief searches of approximately three words are typical, unless the searchers refine their contextual understanding by gathering additional information about their topics. Although reflexive searching is usually effective enough for simple information needs, greater awareness and planning are necessary for more complex questions.

The literature further hints at these issues via studies such as the one by Monroe-Gulick and Petr (2012), in which students claimed to experience difficulty with "information saturation" (329). In Leibiger's (2011) article on the prevalence of "Googlitis" in college classrooms, she positions Google as a rhetorical proxy for the lack of reflective search among library users, and she moves students away from Google products through targeted instruction sessions. Key to her argument is that Google searches are "unsophisticated" (189) in comparison to the robustness of search in subject databases (195). Kingsley et al. (2011) worked to provide more concrete evidence of Google and Google Scholar use in the academy when they looked at the information literacy skills of dental health students. Many of these students started with Google—40 percent listing it as their top choice, just ahead of *Wikipedia* (4). The researchers intervened by teaching students to use PubMed, with a marked improvement in their information literacy skills at the session's end.

Only a handful of articles suggest the value of teaching Google Images, and several more relate Google Images to visual literacy. In their study of the image-searching habits of undergraduates, Bridges and Edmunson-Morton (2011) found that two-thirds of their survey participants used Google or Google Images as their starting point for digital images; one-third of the survey participants *only* used Google or Google Images (29). For these authors, "dismissing Google, when it has proven successful for patrons in the past, only serves to undermine the librarian's role in the research process" (30). In our "Finding Images" workshop, we propose some database alternatives, but communicate that Google tools can have a place in academic research. This provides a tremendous opportunity to improve visual literacy skills while respecting researchers' established workflows.

Google creates an information hunger in our attendees. Learning that librarians know their most-used search engine for images in addition to specialized databases prompts them to ask further questions about search and their relationship to it. The reception of the workshop, along with the changes we made, shows the moving target of this gradual awakening to visual literacy. We have had fruitful discussions with attendees about specific reference questions following our workshops.

However, the Google search team's frequent changes to the platform require constant vigilance—and our workshop has allowed us to alert participants to this. Database vendors inform librarians via e-mail about platform revisions; if one keeps up with search industry news, Google provides the same notifications to everyone. News outlets inform readers about changes to Google that might impact them, and Google has several blogs that warn of changes, such as the Google Blog (http://googleblog.blogspot.com/) and the Inside Search Blog (http://insidesearch.blogspot.com/). There is a caveat: Once, we had a short planning session the day before the workshop and awoke the next morning to find that Google had pushed search changes requiring a last-minute handout revision, and they provided little advance warning. We included this anecdote in the session to illustrate how quickly and covertly Google transforms.

A story in *Forbes* from March 2014 promises "big changes" to the way we see search results (DeMers 2014). These include a more ubiquitous use of the Knowledge Graph included as a scrolling bar at the top of many search results, a greater emphasis on expert content, and increased personalization of search results. The Knowledge Graph already appears in selected Google Images searches. For example, a search for *jaguar* retrieves a Knowledge Graph with links to images of Jaguar cars, jaguar cubs, jaguar vs. leopard, the Jaguar sports car logo, and images of a popular model of Jaguar vehicle. A user can scroll through this list independently of the general search results and click on the relevant jaguar topic to view all of the results for their new search. We can use this feature to highlight the benefit of subject search filters for images across all image resources, not just Google Images, but the personalization of search results may make it difficult to ensure that our attendees have a uniform workshop experience.

FINAL THOUGHTS

As librarians, we seek to enhance the overall research experience by meeting people where they are. We have integrated Google's products with subscription databases and curated digital collections to break down barriers between familiar and esoteric tools, with the goal of promoting self-reflection in search. Google Images offers a powerful starting point for conversations about many aspects of visual literacy, such as an image's provenance and copyright restrictions. It can also serve as a complement and a contrast to other fee and subscription-based image resources. Because visual literacy instruction is an area of emerging best practices, a workshop on Google Images at another institution might evolve differently.

Now that we have successfully found the appropriate ratio of Google Images to other workshop content, we continue to hone this session to meet the needs of our community. This includes our decision to involve a copyright librarian in the next iteration of the workshop who can speak authoritatively about fair use and copyright clearance of images. With the introduction of our colleague's expertise, we realize that our workshop is no longer just about *finding* images—and so, our next session will be called "Finding and Using Images for Scientists and Social Scientists."

REFERENCES

Bridges, Laurie M., and Tiah Edmunson-Morton. 2011. "Image-Seeking Preferences among Undergraduate Novice Researchers." *Evidence Based Library and Information Practice* 6(1): 24–40.

Choi, Youngok. 2010. "Effects of Contextual Factors on Image Searching on the Web." *Journal of the American Society for Information Science and Technology* 61: 2011–28. doi:10.1002/asi.21386.

DeMers, Jayson. 2014. "Big Changes Are Coming to Google Search Results—Are You Ready? [Web Log Entry]." *Forbes Online*, March 6. www.forbes.com/sites/jaysondemers/2014/03/06/big-changes-are-coming-to-google-search-results-are-you-ready/ .

Hattwig, Denise, Joanna Burgess, Kaila Bussert, and Ann Medaille. 2011. "ACRL Visual Literacy Competency Standards for Higher Education." www.ala.org/acrl/standards/visualliteracy .

Hattwig, Denise, Kaila Bussert, Ann Medaille, and Joanna Burgess. 2013. "Visual Literacy Standards in Higher Education: New Opportunities for Libraries and Student Learning." *Portal: Libraries and the Academy* 13(1): 61–89.

Kingsley, Karl, Gillian M. Galbraith, Matthew Herring, Eva Stowers, Tanis Stewart, and Karla V. Kingsley. 2011. "Why Not Just Google It? An Assessment of Information Literacy Skills in a Biomedical Science Curriculum." *BMC Medical Education* 11(1): 17.

Leibiger, Carol A. 2011. "'Google Reigns Triumphant'? Stemming the Tide of Googlitis via Collaborative, Situated Information Literacy Instruction." *Behavioral & Social Sciences Librarian* 30(4) (October): 187–222. doi:10.1080/01639269.2011.628886.

Mikkelson, Barbara, and David Mikkelson. 2012. "The Imperfect Storm." *Snopes.com*. www.snopes.com/photos/natural/nystorm.asp .

Monroe-Gulick, Amalia, and Julie Petr. 2012. "Incoming Graduate Students in the Social Sciences: How Much Do They Really Know about Library Research?" *Portal: Libraries and the Academy* 12(3): 315–35. doi:10.1353/pla.2012.0032.

Chapter Ten

Enhancing Information Literacy Instruction with Google Drive

Laksamee Putnam

Within academia, library instruction can involve lessons on keyword creation, proper database usage, or understanding plagiarism. These lessons often take place in a computer lab allowing students to move from lecture to practical application. For example, they can test if their keywords work or explore various databases. In addition, library instruction should help move students beyond practical skills to developing critical thinking and motivate students to become truly information literate in any situation, whether it is for academic research, everyday life, or their future careers. However, academic library instruction usually means a single one-shot session within a semester-long class. Realistically, making the most of minimal time can mean librarians are more focused on convincing students the library is a useful and reliable resource (LaGuardia 2012). It can be easy for librarians to lose sight of the bigger picture as we struggle to adapt to a decentralized information environment. The ACRL's Information Literacy Competency Standards for Higher Education, approved in 2000, are being revised, in part, because the standards do not "recognize students as content creators as well as consumers and evaluators and do not address ongoing challenges with student learning in a multi-faceted, multi-format, media-rich environment" ("The Future of the Standards" 2014). There is a push within academia to "connect the dots" and move from learning singular skills toward a continuum of education. One such way to encourage a collaborative learning environment and engage students, faculty, and librarians is through the use of Google Drive.

Google Drive (www.google.com/drive/index.html) provides a toolbox of cloud-based, real-time programs that can allow librarians to digitize numerous aspects of their instruction. Integrating Google Drive programs such as

Google Forms or Google Sheets can help librarians integrate peer-to-peer instruction, guide the critical thinking process, and provide tangible evidence of the need for information literacy instruction. Technology and online environments hold the potential to improve cooperative learning, and their ubiquity has made access possible for most students. However, it is not necessary for instruction librarians to reinvent the wheel in order to innovate in their classrooms (Koury and Jardine 2013). This chapter will highlight the benefits and challenges of digital worksheets used for a variety of library instruction sessions, all created with Google Drive.

BACKGROUND

Google Drive is a file storage and synchronization service that allows users with a Google account to upload files, collaborate on projects, and access and edit those items from a variety of devices such as computers, smartphones, and tablets. The productivity applications are similar to Microsoft Office applications, allowing users to create documents, spreadsheets, forms, and presentations. A single account user has the ability to create any of these files and then share with other users (additional accounts not required). The documents can then be edited in real time; any user logged in to the document at the same time can observe changes as they occur. After creating items in Google Drive, users can also send out the documents to interact with an audience. For example, a user creating a survey can construct a fairly comprehensive form with various question types including multiple choice, short answer, checkboxes, and scales. When the form is shared, users' answers are automatically filled into a spreadsheet, and the creator can make the results accessible, or keep them private.

It is worth noting that Google Drive was created in 2012, adapted from a previous application, Google Docs, which had been around since 2007. Along the way, Google has fairly regularly made updates, added new features, and changed the interface, which have been documented at Google Drive Blog (http://googledrive.blogspot.com/). My experience with these changes has been positive, with most updates improving features such as word count, increasing mobility by creating more access points to the applications, and improving the tools through the use of add-ons, for example, a thesaurus. The following classroom examples, therefore, might be outdated by the time the reader takes a look at them. The activities are adaptable to any synchronous file sharing service (Dropbox, iCloud), but the best way to stay current is to choose a service, stick with it, and be ready to change along with the tide of technology updates.

GOOGLE SHEETS: INTERACTIVE WORKSHEETS TO ENCOURAGE HANDS-ON PRACTICE

The majority of digital worksheets I create for library instruction utilize Google Sheets. The spreadsheet is made available to the class through a simple URL and students are asked to follow a lecture and then fill in a row from the various instructions provided across the top of each column (figure 10.1).

The class structure is typically a series of brief lectures followed by in-class work repeated until completion. This incremental process allows me to clarify each step and encourages students to discuss their research method. Students first fill out the spreadsheet with their name and research topic (better if students have already chosen their research topics before the session). This initial step is important, as it allows the students to view the spreadsheet and understand how to enter their work. Students frequently need time to claim a row as their own, and understand that as other students input responses, the worksheet updates. It also establishes time to troubleshoot any students with technical issues. Problems could include the need to use a different browser; frequently an outdated version of Internet Explorer prevents the spreadsheet from working properly. Google Chrome is the recommended browser choice. Another frequent problem occurs when the privacy settings of the spreadsheet are not adjusted to allow students to view and edit. When creating Google Drive documents always double-check the share settings to be sure the correct audience has access and permissions to edit.

Choose a row below and Enter Your Name	**Enter Your Topic**	**Form your first keyword/synonym into a search strategy**	**Form your second keyword/synonym into a search strategy**	**Form your third keyword/synonym into a search strategy**	**Find a website resource for your topic using Google Advanced, Cook's Guide to the Web, or Duck Duck Go. Enter the URL for the website you found**
Example: Laksamee Putnam	Which age is childhood obesity in the United States the highest?	obese OR obesity OR overweight	child* OR youth OR adolescent	"United States" OR USA	http://www.aacap.org/cs/root/facts_for_families/obesity_in_children_and_teens
	Funding in public schools	fund* OR budget OR money	"public schools" OR education	"United Sates" OR USA	http://educationnext.org/public-schools-and-money/
	To what extent should funding be increased or decreased for the arts in public educational systems.	US OR AmericaORr United States	Arts OR Dance OR Theatre	funding OR money OR finances	
	Effects of dugs and alcohol effects on students	drugs and alcohol	teenagers OR students OR youth	cons and negative effects	http://www.drugabuse.gov/publications/drugfacts/high-school-youth-trends
	Social Media: good or bad?	social media OR twitter OR facebook OR instagram OR United States	good OR bad AND pros and cons	age group OR children OR adolescents OR youth	http://pediatrics.aappublications.org/content/127/4/800.full

Figure 10.1. Sample spreadsheet from an instruction session. Please view the full spreadsheet here: http://bit.ly/spreadsheetsample

After establishing a structure for the students to enter their work into the spreadsheet, there are several different ways to organize the column instructions. In figure 10.1, students are being asked to break their research topic into separate keywords and then use Boolean search terms to create a search strategy. Providing an example row across the top allowed me to demonstrate what I expected their answers to look like. As the students work, they are motivated to keep up, as they view peer rows being filled in. Typically, I have observed a cascade effect where a few students fill in thorough answers, sparking the thought process of other students. Pausing student work to discuss various entries and clarify common questions also allows the instructor to actively correct common mistakes and assess the students' level of understanding. While this is possible with standard paper worksheets, the digital spreadsheet creates an environment that showcases all the student work in a single location, speeding up the assessment process and perhaps reaching students who do not ask direct questions, or do not realize they are making mistakes. A colleague working with a similar digital worksheet observed that spreadsheets benefit the classroom by creating transparency, allowing the instructor to provide feedback during class rather than adjusting the curriculum afterward (Simpson 2012).

After filling in keywords, the students are asked to search for a website resource on their topic and evaluate for currency, reliability, authority, and purpose. The lecture I provide on evaluating sources describes the criteria by highlighting a number of hoax websites, such as the "Help Save the Pacific-Northwest Tree Octopus" site (http://zapatopi.net/treeoctopus/) , to spark student interest. As students fill in their website resources and evaluations, discussions frequently arise from a website missing critical information. The spreadsheet provides easy access to the URL the student has entered, so the entire class can analyze the website. Often, these discussions center around student uncertainty over an information format, prompting the class to deliberate over a brochure, a website, or an embedded YouTube video, but these conversations can still revolve around evaluation. Librarians can touch on a number of relevant questions as long as they are prepared to briefly deviate from planned content. Allowing these examples to be discussed as they are discovered provides ample evidence for the need to critically think about the source of information while also integrating new media formats students often uncover while navigating a search. In addition to convincing students to evaluate their information, the spreadsheet also serves as a showcase for faculty, providing data on their students' information literacy skills. If the faculty is present during the library session, the spreadsheet can help them know when to step in to help clarify the research needs for the class and emphasize the importance of learning research skills. Also, after viewing the student work, faculty often realize why library sessions beyond basic bibliographic instruction are so important.

One example of moving beyond bibliographic instruction can be found in figure 10.2. This spreadsheet asks students to use a specific database to find a book and an article on their research topic. Students are placed into small groups and asked to critique the database while finding the book and article. Instead of beginning the session with a lecture pointing out the various features of a database, I flipped the class and integrated peer instruction. Prior to class, students were asked to watch two instructional videos outlining how to locate books and articles; class time then focused on practicing those skills. Instead of a lecture about how to use the database, students worked together on their own. Basic information literacy instruction benefits from peer-assisted learning by allowing students to use language they understand to describe processes librarians are prone to overcomplicate (Bodemer 2014).

As the spreadsheet is filled, I moved between groups, clarifying questions about call numbers, interlibrary loan requests, and other common issues. Once groups have had a chance to reflect and critique the database, they are asked to present their groups' thoughts. With the contents of the spreadsheet available I was able to quickly skim and prepare for each presentation. As groups bring up various features that aided their search, or point out problems they had, the critiques can be used as a jumping-off point for teaching research skills. This flexible class structure can potentially backfire; with discovery services improving, often students can enter their search terms and all the results seem relevant, leaving the impression that the entire search process is easy. However, with the students in groups and with multiple

Enter your Group Name and Members	Choose one of the topics below	Find a book available at Towson University on your topic. List the Title, call number and location.	Find a book available at another USMAI library on your topic. List the title and location.	Enter good/bad features of OneSearch which helped your team find Books	Stop and report back
Team Example: Laksamee Putnam	Childhood obesity	Encyclopedia of contemporary American social issues / HN59.2 .E343 2011eb / Reference ebook electronic access	Our overweight children : what parents, schools, and communities can do to control the fatness epidemic / University of Baltimore	Enter good/bad features of OneSearch which helped your team find Books	Stop and report back
	Peer Pressure	Youth peer-to-peer groups influence attitudes and behaviors of teenagers	Peer pressure : deal with it without losing your cool. Coppin University	The FORMAT feature, LOCATION feature	
	cyberbullying	Cyberbullying : activities to help children and teens to stay safe in a texting, twittering, social networking world/Shady Grove Library Stacks/ HV6773 .R64 2010	Cyberbulling; Preventio and Response/ Towson University E-Books	location, call number, and format	
	Durgs and alcohol effects on students	Secondary effects on heavy drinking on campus	College drinking and drug use/ edited by helene rasking white, david l. rabiner/ Coppin University	Location, the summary, fromat, number to call	
	public school funding	Money, mandates, and local control in American public education/ Bryan Shelly/ Towson University E-books		Bad: Found a book yesterday but cannot find it today. Good: the book I found fits perfectly with my topic	

Figure 10.2. Sample spreadsheet showing student group work. Please view the full spreadsheet here: http://bit.ly/spreadsheetsample2

groups all entering data onto a single spreadsheet, it quickly becomes clear that research is not a straightforward process. Class discussions can create a cohesive lecture, as long as the instructor is prepared to interject with questions about elements that might be missing from student critiques. I found most students struggled with a large number of results but a lack of relevance; this situation mirrors current research, which points blame toward the globalization of information and a shift to Googlized search results (Fu and Thomes 2014). Frustration with this part of the research process can cause students to give up. The spreadsheet activity allows students to peer-teach the basic bibliographic skills, but also provides the instructor with a way to teach advanced database skills from a point-of-need perspective. Overall, students can begin to comprehend the need to practice using databases and how to narrow results in order to find the right content.

GOOGLE FORMS: ASSESSING STUDENT KNOWLEDGE

Google Forms is a survey application. There are numerous free survey tools available, including Typeform, Survey Monkey, and Poll Daddy. Additionally, more advanced features and improved analytics can usually be purchased through most online survey tools. Google Forms is part of the Google Drive suite and does not offer any premium features. Consequently, Google Forms is not as robust as most of the other tools on the market. However, for simple information gathering, Google Forms is a convenient way to collect basic data.

I use Google Forms as a way to gather pre- and post-class evaluations. Understanding what students know, or think they know, can help the instructor prepare session content. After class, hearing what students enjoyed, or disliked, can benefit the next iteration of that class. Since most academic library instruction occurs as an integrated part of a semester-long course, course evaluations often do not specifically survey student satisfaction with the library session. Building in your own evaluations can be a useful way to showcase library value and support the need for library instruction.

A slightly more complex use of Google Forms can be found in figure 10.3. After a lecture detailing various citation rules, students are asked to copy the book and article reference from a built-in database citation creator and then correct the mistakes.

After submitting their answers, the correct versions of the references are shown, and any student questions are addressed. Unfortunately, the form limits the ability to format the entry (e.g., italics). This may be a feature that will be added to Google Forms later; however, with some features not available, readers may want to explore other survey options if their needs are complex. Instructors who create a form can view all the responses in an

APA Citation Practice

A proper citation/reference makes it possible for someone to find the source of your information. There are style guides available on the Albert S. Cook Library website (http://cooklibrary.towson.edu/styleGuides.cfm). Feel free to use these to help you as you create references for the various books and articles below.

Enter your name

Books

A basic print book reference requires:

1. Author(s)'s last name(s) and first initial(s) (if known)
2. Year of publication
3. Title of book/monograph
4. Publication information: City, state abbreviation: publisher's name.

An example book citation (note that this form prevents me from including proper formating such as italics):
Hemingway, E. (1964). A moveable feast. New York, NY: Scribner's

Create an APA reference for http://bit.ly/VZpYiU "Education nation: six leading edges of innovation in our schools"
Don't worry about the hanging indent. Please put "quotes" around what needs to be italics! Ask if you have questions!

Figure 10.3. APA citation practice. Please view the full form here: http://bit.ly/apaformsample

online spreadsheet, simplifying the grading process. It also makes recognizing trends in the answers easier, clarifying what might need to be emphasized. For example, if most students are forgetting to convert author first names to initials, this concept can be quickly brought to the class's attention. Whether using Google Forms to evaluate instruction or to gather student work, the application provides an innovative way to digitize instruction, especially in a computer lab environment where students are sometimes more focused on a monitor than the teacher.

CONCLUSION

Google forms and Google spreadsheets are both simple to make and easy to integrate into a library instruction session. The activities shown in this chapter enabled collaborative learning by making library instruction less lecture based and more active. Higher education is embracing the merits of collaborative technology, especially as research has shown it to increase student learning (Zhou, Simpson, and Domizi 2012). Information literacy skills are an integral part of critical thinking, and students can best learn those lessons when faculty and librarians work closely together. Google Drive can be a great way to show faculty how to embed librarians into the curriculum (Midler 2012). Additionally, it can open the way for integrating other technologies and teaching styles. The lessons here began with Google Drive replacing paper worksheets; however, through the past two years, the lessons have evolved into a flipped classroom where students watch basic instruction videos or read various articles before class, and then come to class prepared to practice. Along with Google Drive, other web-based technologies such as Poll Everywhere and SlideShare are used to benefit the classroom by including universal design aspects, addressing multiple learning styles, and increasing accessibility. As the technology improves, or better collaborative tools are designed, librarians must be ready to adapt their teaching. For example, Google Drive also includes a Google Slides application, which allows users to create a basic presentation similar to Microsoft PowerPoint. However, because Google Drive is integrated with other various Google resources, a user can search within the Google Slides interface for pictures or possible references to pull into the presentation. An example lesson plan integrating this feature can be found on the Free Technology for Teachers website, which compares Google slides to index cards used to help students study or organize their thoughts (Byrne 2014). More advanced integration may be possible; perhaps in addition to Google results, the search could also include local library resources or subscription access to peer-reviewed materials, making research more seamless. This also emphasizes the importance of teaching critical thinking skills, preventing students from accepting results out of convenience.

Of course, just because a new technology becomes available does not mean it is suited for educational purposes. While library instruction should be adaptable, it remains important to focus on learning outcomes rather than using the new technology just because it is available. Using technology "properly" to teach has moved many disciplines away from lecture, and brought hands-on activities to the forefront of positive pedagogy. Information literacy instruction can follow this same path, adapting to help students navigate today's media rich environment. The use of Google Drive is just one example of how a librarian can increase class participation and also help students develop vital research skills.

REFERENCES

Bodemer, Brett B. 2014. "They Can and They Should: Undergraduates Providing Peer Reference and Instruction." *College & Research Libraries* 75(2): 162–78. http://crl.acrl.org/content/75/2/162.full.pdf+html .

Byrne, Richard. 2014. "How to Use Google Slides to Organize Research." *Free Technology for Teachers*. http://www.freetech4teachers.com/2014/02/using-google-slides-to-organize-research.html#.U3-yb_ldWSo.

Fu, Li, and Cynthia Thomes. 2014. "Implementing Discipline-Specific Searches in EBSCO Discovery Service." *New Library World* 115(3): 102–15. doi: 10.1108/NLW-01-2014-0003.

"The Future of the Standards." 2014. *Framework for Information Literacy for Higher Education.* Accessed May 27, 2014. http://acrl.ala.org/ilstandards/?page_id=19.

Koury, Regina, and Spencer J. Jardine. 2013. "Library Instruction in a Cloud: Perspectives from the Trenches." *OCLC System & Services* 29(3): 161–69. doi: 10.1108/OCLC-01-2013-0001.

LaGuardia, Cheryl. 2012. "Library Instruction in the Digital Age." *Journal of Library Administration* 52(6): 601–8. doi: 10.1080/01930826.2012.707956.

Midler, Zoe. 2012. "Case Profile: Zoe Midler and Google Docs." *Library Technology Reports: Embedded Librarianship: Tools and Practices* 48(2): 12–15. http://www.alatechsource.org/ltr/index .

Simpson, Shannon R. 2012. "Google Spreadsheets and Real-Time Assessment: Instant Feedback for Library Instruction." *College & Research Libraries News* 73(9): 528–30, 549. http://crln.acrl.org/content/73/9/528.full .

Zhou, Wenyi, Elizabeth Simpson, and Denise Pinette Domizi. 2012. "Google Docs in an Out-of-Class Collaborative Writing Activity." *International Journal of Teaching & Learning in Higher Education* 24(3): 359–75. http://www.isetl.org/ijthe .

Chapter Eleven

Fusion Tables for Librarians and Patrons

Rebecca Freeman

Data can be cumbersome to work with, especially large amounts of data. There are options out there to analyze this data, but many of these are expensive. This how-to chapter will explore ways to use the experimental application Google Fusion Tables, the pros and cons, and will give you step-by-step directions for creating your own Google Fusion Tables chart.

Fusion Tables is a new application available through Google Drive that utilizes data in a spreadsheet format. It works with Google Sheets, Excel, and Calc (an Open Office software application much like Excel). Fusion Tables allows the user to visualize data through a variety of means such as charts.

WHY USE GOOGLE FUSION TABLES

I became aware of Google Fusion Tables when our campus Academic Success Center needed a way to create reports. The Academic Success Center houses tutors for multiple disciplines. They needed a cost-effective way to run reports from data collected by their tutors; Google Fusion Tables fit the bill perfectly.

Since then, I have used Google Fusion Tables for reports, statistics, and research. It can be used by librarians and patrons alike in a variety of ways, including:

1. Collection development: Determine the age of your collection, collection gap identification
2. Research: Analyzing data to determine if more cookbooks in specific styles are written by men or women

3. Statistics: User statistics, area use statistics, furniture use statistics
4. Mapping: Analyzing data to determine where different species of butterflies live

PROS AND CONS

Google Fusion Tables is useful because it is versatile and customizable. The user determines how the data will be used and analyzed and in which format to present the data. It has been useful for Medford Library in part because it is free and is easy to start using. Fusion Tables also allows for multiple views of the data at the same time through the use of different tabs. Each tab can hold a different format of the data, which can be filtered in different ways.

Using Google Fusion Tables does require that the user has a Google account. Fusion Tables is an application that is available through Google Drive, which is part of a Google account. While the user must have a Google account, the Fusion Tables can easily be shared with those who do not have an account.

While Google Fusion Tables is very useful, there are still some glitches, due in part to the fact that it is an experimental application. While one of the advantages to Google Fusion Tables is that it is free, this means that the user will need to do some additional work that a more costly product might automatically do.

There are also some things that need to be kept in mind when using Google Fusion Tables, including:

1. It is not safe to put Social Security numbers or other protected information into a Fusion Table.
2. Ensure that data used is clean and corrected prior to placing it in a Fusion Table (there is no way to get rid of duplicate data).
3. Charts and graphs do not always work well with large amounts of data. (Currently there are no clear parameters for the amount of data it can handle. The application will simply give an error message without explanation, and that can be confusing to users.)
4. Google Fusion Tables only recognizes one data point and may not recognize headers or spaces between data.

TERMINOLOGY

Google Fusion Tables has its own language, and it's useful to have a basic understanding prior to starting.

Rows: The row format is the default and contains all of the data for a specific data point. It looks like what users see when using any spreadsheet.

Card: The card format displays the data in cards that look like business cards. This format can be customized to display what data the user wants.

Summary: The summary format creates an overview of the data.

Chart: The chart format can be customized for a variety of charts, for example, pie chart, bar chart, or scatter plot. It is customizable, allowing the user to change the colors, add a title, add labels, make it 3D, as well as many other custom features.

Mapping: The mapping feature uses the name of a location or coordinates for the location to place it on a Google Map.

Filter: The filter uses the headers to narrow down data and is available for use in any of the formats.

Criteria: When using Google Fusion Tables there are some criteria for what format the data can be in and how large it can be.

Google Fusion Tables allows for a spreadsheet in any of the following formats to be imported:

.csv
.tsv
.kml
.xls
.xslx
.ods

The size of the spreadsheet can also affect the performance. Spreadsheets must not exceed these limits:

100 MB
1,000 columns per table
1 MB of content per row

ADDING GOOGLE FUSION TABLES

To start using Google Fusion Tables it must first be added to the user's Google Drive. Follow these directions to add Google Fusion Tables.

1. Log in to your Google account.
2. Go to Google Drive.
3. Select Create.
4. Select Create More Apps listed on the bottom of the drop-down menu.

5. Once the new window opens, search for Fusion Table in the search box.
6. *Fusion Table (experimental)* will appear.
7. Select Connect.
8. A message will show up asking whether Google Fusion Tables should be the default app for the files it can open.
9. Select the correct option for that specific user.
10. Click OK.

Google Fusion Table is now attached to the user's Google Drive account. Adding Fusion Tables only needs to be done once.

CREATING A GOOGLE FUSION TABLE

First Time Use

The first time Google Fusion Tables is opened it will require that the user reenter their password.

1. Select Create.
2. Select Fusion Table (experimental).
3. Reenter Google password.

Create a Table from a Spreadsheet on Your Computer

When working with data, frequently that data is in the form of a spreadsheet that is housed on a computer. Here is how to get that information from the computer to a Fusion Table.

1. Select Create.
2. Select Fusion Table (experimental).
3. Select From Computer.
4. Select Choose File.
5. A search box will appear.
6. Pull up the correct file on the computer.
7. Highlight the correct file.
8. Click Next.
9. A screen will pop up asking which row is the header.
10. The default is the first row to be the header.
11. Any rows above the header will not show up in the table.
12. Click Next.
13. Name the table (it is possible to rename the table later if needed).

14. Check whether the data can be exported by others or not (the default is that the data can be exported by others).
15. Attribute the data by filling out where and when it was obtained (this is important if you did not collect the data).
16. Enter a description of the table (this should be something that will help you remember what the table is about if you need to access it later).
17. Select Finish.

See "Using Data" below for steps on how to use the data in the table you just created.

Create a Table from a Google Sheet

It is possible to create a table from data in a Google spreadsheet. This is useful when working with data gathered from a Google form. These steps will walk you through creating a Fusion Table from a Google spreadsheet found on your Google Drive account.

1. Select Create.
2. Select Fusion Table (experimental).
3. Select Google Sheet.
4. All of the Google Sheets on your account will appear.
5. Click on the spreadsheet you want to work with.
6. A check will appear in the box on the left-hand corner of the spreadsheet you've selected.
7. Click on Select. (Short cut: Double-click on the picture of the correct spreadsheet to select instead of clicking on Select.)
8. A screen will pop up asking which row is the header.
9. Any rows above the header will not show up in the table.
10. Click Next.
11. Name the table (it is possible to rename the table later if needed).
12. Choose whether the data can be exported by others or not (the default is that the data can be exported by others).
13. Attribute the data by filling out where and when it was obtained (this is important if you did not collect the data).
14. Enter a description of the table (this should be something that will help you remember what the table is about if you need to access it later).
15. Select Finish.

See "Using Data" below for steps on how to use the data in the table you just created.

CREATE AN EMPTY TABLE

An empty table is another type of table you can create. This table creation method requires the user to provide the data either imported or input manually. To create an empty table:

1. Select Create.
2. Select Fusion Table (experimental).
3. Select Create Empty Table.
4. An empty table will appear.

See "Making Changes to a Row" for directions on how to add data to a row. See "Adding More Data" for directions on how to place data into the table.

Renaming Headers

Empty tables have default headers: Text, Number, Location, and Date. When adding data to these it may become necessary to change the headers. To rename headers:

1. Open the correct Fusion Table.
2. Place the mouse over the default header that needs to be changed and an arrow for a drop-down menu will appear.
3. Click on the drop-down menu.
4. Click on Change.
5. This will pull up a box, Change Column.
6. Enter the name of the column (this will be the header name).
7. Enter a description of the column.
8. Select the type of column (e.g., you will be entering text, numbers, or the date into this column).
9. Select the format (e.g., you will be entering pictures, YouTube videos, etc.).
10. Select whether the data will be validated.
11. If you want to create a drop-down box for the column, list the items that will be available in the list.
12. Select Save.

Adding a New Column

When working with an empty table or even a table with data in it, the user might find that it is necessary to have an additional column. This will add additional data to each row. Adding this additional data will need to be done manually, see "Manual Data Entry" below for more details. To add a new column:

1. Open the correct Fusion Table.
2. On the tool bar, select Edit.
3. Select Add Column.
4. A box, Add Column, will appear.
5. Enter the name of the column (this will be the header).
6. Enter a description of the column.
7. Select the type of column (e.g., you will be entering text, numbers, or the date into this column).
8. Select the format (e.g., you will be entering pictures, YouTube videos, etc.).
9. Select whether the data will be validated.
10. If you want to create a drop-down box for the column, list the items that will be available in the list.
11. On the left of the box is a list of the column in your table.
12. When the name is highlighted, an up arrow, a down arrow, and an *x* will appear.
13. Determine the location of the column by moving it up or down the list with the up and down arrows.
14. Select Save.

MAKING CHANGES TO A ROW

There are many reasons that it might be necessary to make changes to data within a row, from just making a spelling correction to adding or updating data. This section will direct you through opening a data point and making corrections.

1. Open the correct Fusion Table.
2. Click on the row that you need to make changes to.
3. Three little icons will appear on the first cell of the row; the pencil allows you to edit, the two pieces of paper allow you to duplicate the row, and the trashcan allows you to delete the row.
4. To edit, select the pencil icon.
5. An Edit Row box will appear with the headers next to the fields for those headers.
6. Enter the data in the correct fields.
7. Click Save if you are done making corrections.
8. Click Save and edit the next if you need to make corrections to the next row of data.

9. Repeat steps 5–7 as many times as necessary
10. On the last row of data you will only have the option to Save.

ADDING MORE DATA

When more data becomes available the ability to add more to a table can be very useful. This section will show you the steps for two ways of adding data: manual data entry and importing more data.

Manual Data Entry

The first way to enter data or add data to a Table is through manually adding it.

1. Open the correct Fusion Table.
2. Go to the tool bar and click Edit.
3. Select Add Row.
4. A box will appear. Add the new row, with the headers next to empty fields.
5. Enter the data into the correct fields.
6. If you are only adding one row, click Save when the data is entered.
7. To add multiple rows, click Save and add another when the data is entered.
8. When adding multiple rows, repeat steps 4 and 6 until you have entered all of the data.

Import More Data

The second way to add more data to a table is to import it from another spreadsheet. Two ways to do this are: importing from your computer and importing from a Google spreadsheet.

To import from your computer:

1. Open the correct Fusion Table.
2. Go to the tool bar and click Edit.
3. Select Import More Rows.
4. Click Choose File.
5. Select the correct file from your computer files.
6. Select Open.
7. The file name should now be listed next to Choose File. (If the correct file is not listed, follow steps 4–6 again to get the correct file.)
8. Select Next.
9. A box will appear titled Import More Rows into New Table.

10. The box will have two sections, the existing column and the new columns.
11. Select the columns in the new file that match the existing columns by checking the box over the column.
12. Select Finish.

To import from a Google spreadsheet:

1. Open the correct Fusion Table.
2. On the tool bar, select Edit.
3. Select Import More Rows.
4. Select Google Sheets.
5. A box will appear with a list of the spreadsheets you have on your Google Drive account.
6. Select the spreadsheet you want to work with by clicking it.
7. A check mark will appear in the box on the left-hand corner of the spreadsheet screen shot.
8. Click Select.
9. A box will appear titled Import More Rows into New Table.
10. The box will have two sections: the existing column and the new column.
11. Select the columns in the new file that match the existing columns by checking the box over the column.
12. Select Finish.

USING THE DATA

The whole point in creating a Fusion Table is to analyze data. There are many different ways that Fusion Tables can be used to analyze data, depending on the data that the user is working with. This section introduces a couple of approaches and will give the user the tools to create and work with the other formats.

Filter

One of the reasons that Google Fusion Tables is such a great tool is that the user can easily filter through large amounts of data. The filter is created automatically with the headers used as the filters. The filter can be applied to any of the layouts and is transferable from one tab to another. To apply a filter:

1. Open the correct Fusion Table.
2. Choose the correct tab to filter.

3. Click on the blue Filter button.
4. A drop-down box will appear listing all of your headers.
5. Select the correct header to filter.
6. Check the boxes for the information you want to view (you can select as many or as few as needed).
7. If the specific value you are looking for does not appear, enter it into the search box under the name of the header and click Find.
8. To add additional filters follow steps 3–5 as many times as needed.
9. The table will automatically update when a value is checked.

Create a Row Layout

As stated before, the row layout is the default for Fusion Tables. This is the format that the data will appear in whether the user creates an empty table or creates one from a spreadsheet. It is necessary in some cases to create a new tab in row format. To create a row layout:

1. Open the correct Fusion Table.
2. To the right of the tabs will be a little red box with a +.
3. Select the +.
4. A drop-down box will appear.
5. Select Add Row Layout

Create a Summary

When using some data, an overview is important to see what is happening within the data. An overview can be created by using the summary layout. To create a summary layout:

1. Open the correct Fusion Table.
2. To the right of the tabs will be a little red box with a +.
3. Select the +.
4. A drop-down box will appear.
5. Select Add Summary.
6. A box will appear. Choose Summary Columns.
7. Choose what the data will be summarized by.
8. Click Add Another if the data is being summarized in multiple ways.
9. Check the box for Bar Charts if you want the data in a bar chart format.
10. Check the box for Summary Count for a summary count of a specific header.
11. Choose which column should be summarized.

12. Under the drop-down box for the column you choose, there are four choices: Minimum, Maximum, Average, and Sum.
13. Check the box next to the ones that you want shown in the bar chart and the table.
14. Select Save.

Create a Pie Chart

The chart that is created with a summary is a bar chart but with some data looking at it in pie chart format makes it easier to understand quickly. To create a pie chart:

1. Open the correct Fusion Table.
2. To the right of the tabs will be a little red box with a +.
3. Select the +.
4. A drop-down box will appear.
5. Select Add Chart.
6. A chart will appear in the middle of the screen with screen shots of other charts listed on the left.
7. Select the screen shot of the pie chart.
8. Choose the correct Category.
9. Check the box next to Summarize Data? if you want a summary count.
10. Choose the Value.
11. Click on the drop-down box under Sort By to choose what column will be used to sort the data.
12. Click Change appearance to update how the graph looks.
13. A Chart Editor box will appear.
14. Add a title.
15. Make changes to the appearance of the chart as necessary.
16. Click OK.
17. Select Done.

Rename a Tab

Each time a new format is created a new tab is created with default and nondescript names (e.g., Summary 4). It can get confusing keeping track of the different tabs. The name of the tabs can be changed to reflect what is being done on each tab. To rename a tab:

1. Open the correct Fusion Table.
2. Click on the correct tab.
3. Click on the down arrow next to the tab name.
4. Select Rename.

5. Enter the correct name for that tab.
6. Select OK.

LIMITED SHARE

One of the advantages to Google Fusion Tables is how easy it is to share your data with collaborators. You can choose how much control those collaborators have over the data. In this example, the Fusion Table will be shared with collaborators so that they can see the information, but they will not be able to change anything or share with others.

1. Open the correct Fusion Table.
2. Select Share in the top right hand corner.
3. A Sharing Settings box will appear.
4. On the very bottom of the box is a message that editors can add people; select the blue Change next to the message.
5. Select Only the Owner Can Change the Permissions.
6. Select Save.
7. Under Invite People, enter the e-mail addresses for your collaborators.
8. To the right of the field with your collaborators' e-mail addresses is a blue drop-down box, Can Edit.
9. Click the drop-down box.
10. Select Can View.
11. Select Send.

PUBLISH

Fusion Tables encourages the easy publication of data. There are a variety of ways to publish this information including embedding a chart on a website. This example will show how to get the code to embed a chart in a website.

1. Open the correct Fusion Table.
2. Select the tab with the correct chart.
3. On the tool bar select Tools.
4. Select Publish.
5. If the table is private, the visibility will need to be changed; see Change Visibility below for more details.
6. Select the HTML code to use to embed into a website.

Change Visibility

The default for the Fusion Tables is that the table is private. You do not need to change this if you are sharing with collaborators, but if the information will be made available to the general public, this setting needs to be changed.

1. Open the correct Fusion Table.
2. Select Share in the top right-hand corner.
3. A Sharing Settings box will appear.
4. Under Who Has Access is a list of who has access and what the privacy setting is.
5. Select the blue Change.
6. Select Anyone with the Link.
7. Select Save.
8. Select Done.

CONCLUSION

Google Fusion Tables is a useful tool for librarians as well as patrons. It allows for analyzing and visualizing big data. Fusion Tables is very versatile in what it can do and the data it can handle. It is loaded with great tools that allow for customization and ease of publishing. This is a free product that is especially for financially strapped institutions and users. As with other free products it does mean more work for the user. Because it is still experimental, there are definite glitches, but overall it fills a void in many areas.

Ultimately, if you are looking for an easy-to-use method of analyzing data and creating charts and maps, Google Fusion Tables is definitely a tool to check out.

Chapter Twelve

Google and Transcultural Competence

Alison Hicks

In 2005, Jean-Noël Jeanneney, then president of the Bibliothèque nationale de France attacked the newly announced Google Books project, arguing that its natural English bias would adversely affect access to and interpretation of the rest of the world's cultural heritage. Coming soon after the Iraq War, Jeanneney's views were dismissed by some as anti-American rhetoric, or as further evidence of the growing gulf between France and the United States (Green 2014). While this example may be dated, it provides an excellent example of the tensions between diversity and globalization. In this way, and in light of Google's increasingly privileged position as an oracle, or source of knowledge that is trusted far more than it should be, perhaps Jeanneney's comments should now make us stop and question how Google affects how we make sense of the world (Halavais 2013, 2). Perhaps, too, this critical examination of Google's structure and function could then help to ensure that comments like Jeanneney's do not cause such defensive reactions in the first place by helping both teacher and student alike to develop the transcultural competences that are increasingly necessary in today's multicultural societies.

This chapter will explore how a critical examination of Google Search can be used in the information literacy (IL) classroom to help develop students' transcultural competence. The chapter will start by defining transcultural competence and its relationship to IL. It will then move to explore research that looks at how our use of Google can affect our expectations and understandings of the world. The chapter will then draw on the author's work with foreign language students to demonstrate how an examination of the political, social, and cultural dimensions of information practices can help students question and learn to use Google in a more critical manner. Although the chapter pulls from research on foreign language students in a university setting, these themes could easily be adapted to focus on the development of global competences in a variety of secondary and tertiary institutions.

TRANSCULTURAL COMPETENCE

Reactions that rely on stereotypes and ignorance, such as those that met Jeanneney's comments, are neither unique, nor unusual. However, while part of the solution seems to lie in the incorporation of global perspectives into higher education, in practice, education often tends to focus rather superficially on different cultural and literary traditions, or the one-way acculturation of international students to host-society norms. Transcultural competence, however, aims to move beyond seeing culture as exotic or assimilatory. Instead, as the Modern Language Association states, it can be defined as an ability to "reflect on the world and [oneself] through the lens of another language and culture"; being able to develop an understanding of different worldviews by seeing oneself as a member of a society that may be foreign to others (2007, 2). In other words, while the study of cultural narratives is still important, it is clear that stories, legal documents, and political rhetoric from different cultures draw upon their own historical and geographic frames of reference. By examining the background realities of these texts, as well as their form and purpose, students can start to consider alternative ways of "seeing, feeling and understanding things" as well as understanding how these viewpoints may be interpreted by other groups and communities (MLA 2007, 3).

In this way, the development of transcultural competence is particularly important both for work and for life. Globalization means that students must be prepared to "live and work in a society that increasingly operates across international borders" (American Council on Education 2012, 3). Global markets, services, and ideas means that it is no longer just business students who must develop the capacities needed to face global challenges. However, the need for a global outlook can also be seen at home. In the most recent United States census, over 55 million people speak a second language, and millions more live within transnational communities and networks (United States Census Bureau 2011). These realities mean that there is a need for students to understand difference between and within societies.

Transcultural competence is equally important for information literacy. New global realities are also characterized by the growing importance of information—at work, in the community, and in our personal lives. Yet, if these worlds are becoming more multicultural, IL, too, must engage with the new challenges and opportunities. However, IL that is focused solely on teaching students purely functional skills cannot begin to develop the capacities needed to operate within these new multicultural societies. Instead, by basing IL on concepts of transcultural competence, librarians can help scaffold a more holistic and thoughtful way of knowing and acting within different information contexts.

In other words, just as advertising and newspapers (among other cultural narratives) show and can be used to help students understand and interpret differences in meaning and worldview, information, too, reflects the shared knowledge and practices of a community. Information cannot be separated from its context; instead, it is a product of local social, political, and economic realities. In this way, by designing IL that reveals these contrasts, students start to understand that information is culturally specific. In turn, this helps them to see that their own practices are culturally driven, thereby opening the door to understanding difference in international contexts and at home.

GOOGLE AND INFORMATION SOCIETIES

Google, which is a core tool for many students, provides an excellent way to explore these issues of transcultural competence. Its dominant role in society means that it is vital that students understand their search process as well as search results, while its perceived universality often obscures an understanding of how the search engine treats cultural difference. A brief overview of research in this area will highlight the rationale for this class as well as providing background for class discussion points.

One of the first issues that affects culturally focused information practices centers on how Google ranking limits access to authentic cultural materials. User studies show that people expect success from the first few results of a search and rarely click past the first page. In this way, Google has to focus on catering to majority interests to maximize perceptions of success, which has the effect of rewarding large, popular sites that have either been established for a longer time, or that can afford to employ search engine optimization experts (Morozov 2013, 147). Culturally relevant websites, which tend to be minority interest as well as newer or unable to game the system, are therefore less frequently featured on the first page of results. This means that top results for cultural topics are more likely to rely on translations or majority viewpoints rather than culturally authentic materials. Sites in a non-English language are further disadvantaged. Studies have shown that U.S. sites are significantly more likely to be indexed than sites from China or Singapore (Jiang 2014). In addition, there is a very low rate of overlap between country versions of Google (e.g., between google.com.mx and google.com), meaning that unless students know about different versions of Google, they are restricting their searches even further (Jiang 2014).

A complicating factor has been the introduction of personalized search. Studies show that previous search history and geographic location affect about 12 percent of searches (Hannak et al. 2013). Personalized search results are therefore automatically constrained to hide or filter information that we or people in the same region don't agree with, which has obvious implications for finding different points of view. This effect can be seen most poignantly in Graham and Zook's work, which shows that a search in Google maps for the English, Arabic, and Hebrew word for restaurant finds completely unique results (2013). While the filter bubble, as it has become to be known, may not be any more selective than offline personalization, it is clear that these trends change our access to the online world. This was most famously demonstrated by Eli Pariser, whose search for "Tahrir Square" returned either links to news reports or travel agencies (Gillespie 2013).

A second major issue centers on how Google's algorithms can promote racial and cultural stereotypes. Google prides itself on the impersonal algorithms that it uses to order information, claiming that this automated process gives it a credibility that transcends typical media biases. However, it is clear that human editorial judgment is used in the construction of the ranking algorithm, through decisions about which factors to include, what weight to assign each factor, and the value of the content (Grimmelmann 2013). In other words, algorithms are socially constructed, representing "a particular knowledge logic, one built on specific presumptions about what knowledge is and how one should identify its most relevant components" (Gillespie 2013). This causes problems in different cultural environments as Google's judgments represent a particular point of view that may fail to consider cultural context, as the Sahara reporters whose work was censored for being too brutal, demonstrates (Morozov 2013).

In turn, these "objective" algorithms mean that often Google actively presents negative stereotypes of cultural identities. Sweeney discovered that "a greater percentage of ads having 'arrest' in ad text appeared for black identifying first names than for white identifying first names" (2013). Umoja Noble (2013) and Baker and Potts (2013) discovered that Google search and Google autocomplete respectively display extremely negative results for minority groups. This becomes especially problematic as these stereotypes appear normal and unavoidable. As such, it is clear that although early users thought the web would help them escape cultural or racial biases, in fact, "online discourse is woven of stereotypical cultural narratives that reinstall precisely those positions" (Baker and Potts 2013, 187). Combatting this discrimination is just as important online.

These issues are compounded by the search engine's perceived universality, which makes it hard to detect the subtle yet potentially damaging effect on our understanding of the world. In this way, IL that centers on revealing and scaffolding cultural differences will prepare students to step out and act in global information societies.

CLASS STRUCTURE

This class was designed to help students develop their transcultural competence through engaging with Google. Originally developed for an advanced undergraduate Spanish writing class, full details are provided here so it can be adapted to meet different needs.

Instructional Purpose

The class centers on scaffolding a critical engagement with Spanish information environments. This was perceived as the most effective way to move students away from a superficial engagement and understanding of difference. In this way, class was structured around questions about the political economies and cultural authority of knowledge, or, in other words, debate about knowledge—what it is and who decides and creates it. These types of questions, while complicated, help break down the perceived universality of information practices while also helping students to understand the background from which Spanish cultural narratives draw. In this way, the research process was seen as the bridge point between cultures.

At the same time, by structuring the class as a way to explore difference between two cultures, neither culture was positioned as exotic or "normal." This approach is far more inclusive in a class that could contain heritage Spanish speakers (defined as students who grew up in Spanish-speaking households) as well as traditional Anglophone students. Similarly, instead of being presented with strategies or checklists for evaluation, the class draws from personal experience. This approach acknowledges prior knowledge, as well as enabling more scope for critical analysis.

Student Learning Outcomes

1. Students will develop a variety of search and evaluation strategies in order to use the everyday and scholarly Spanish sources that are most appropriate for their research question.
2. Students will demonstrate an awareness of Spanish research practices across different communities, including how each contextualizes and produces information, and how these differ from English-language practices.

Step-by-Step Instructions

The librarian's involvement with the class is centered on one in-class seminar, and three reflective surveys. The in-person seminar is designed to focus on discussion and hands-on practice, with most of the time dedicated to engaging students in questions about the differences between Spanish and English research, and generating resources and strategies designed to help students in their research process. The reflective surveys inform teaching needs as well as student learning needs, and draw upon the work of Troy Swanson, who is one of the few people to provide explicit examples of this type of IL in the classroom (2010). The first survey is designed to help students start to think about their research project while also assessing student needs and prior knowledge. The second survey helps students continue the reflection process by thinking about their learning, while enabling the librarian to assess application of concepts. The third survey is completed at the end of the semester to enable students to reflect on their process after a whole semester of work, while also gauging retention of these concepts.

1. *First Reflective Survey: At least one week before class, administer survey:*

    - What do you already know about your research topic?
    - What do you need to know about your topic?
    - Where might you discover this information?
    - If you wanted to find information about, for example, solid waste in Guatemala, where would you look?
    - When you use Google, how would you look for information in Spanish?

2. *Class Activity: Students will work in groups to find a Spanish resource through Google and record it in this online form. As you work in your group, consider the following questions:*

    - How did you find this resource? How did this differ from English searching?
    - What keywords did you choose?
    - When you were looking at the list of results, what type of resources did you find? What didn't you find?
    - Who can publish on a specific topic?
    - Who can't? And why?
    - What tips do you have for your classmates? (e.g., a great web page, a search tip, etc.)

- Why did you choose that resource? How did you evaluate the results?

3. After each group has found at least one resource (10–15 minutes), start the class discussion about what they found and strategies they employed. Build up a list of search strategies for the class, and a list of criteria to evaluate resources. Probe students for their thoughts and experiences about the more critical questions. Repeat process using the library web page (if appropriate).
4. *Second Reflective Survey: No more than one week after class, administer the second survey:*

   - What changes did you make to your initial searches in order to improve results?
   - What prompted you to make those changes?
   - What are some of your research challenges? How have you dealt with them or what do you need help with?
   - Thinking about the information sources you have found so far, what information do you trust? What causes you to disagree with a piece of information?

5. *Third Reflective Survey: A couple of weeks before the end of semester, administer the third survey:*

   - Who can publish on a specific issue? Who cannot and why? Whose voice is included/excluded?
   - What information is trusted by society? Do you agree?
   - What takeaways from this project or process will you use in your future career or studies?
   - You have been offered a position teaching English in Costa Rica. You have two sessions to teach students who are about to study abroad how to conduct research in the United States. What points will you emphasize, knowing what you know about the differences between Spanish and English research?

This program requires one class session (60–90 minutes) and additional homework time out of class. Activities could be spread over more classes if the students needed more structure.

RESULTS

Various research studies demonstrate that before a research class, Spanish majors have little knowledge of how their usage of Google affects their search results. In a pretest of thirty-two students enrolled in a basic Spanish writing class at the University of Colorado, Boulder (UCB) less than 25 percent of students knew that different country versions of Google exist, while just under half knew that Google's Advanced Search could be used to change the language or regional settings (Hicks forthcoming [a]). In a study of students enrolled in an advanced Spanish writing class at UCB, a pretest showed that although most students recalled at least one way to find Spanish results in Google, a sizeable majority also indicated that they would expect to find relevant Spanish materials in sources that they use for English research papers, for example, JSTOR. (Hicks forthcoming [b]). This data bears out findings from Project Information Literacy that highlights how students rely on the same set of tried and tested resources (Purdy 2012). It also validates the need for this class.

After the class, however, students showed a far greater awareness of strategies they could use to find more culturally relevant results in Google. Furthermore, students demonstrated that they valued these lessons, with over half of students in the basic Spanish writing class indicating that their greatest takeaway from the class was either learning how to use Google or finding materials in Spanish more generally. Students also showed an impressive grasp of the differences between Spanish and English research, being able to reflect on, for example, the lower visibility of Spanish in traditional information systems:

> *Para mi, una investigación española requiere que uno busca más en el internet que los otras fuentes de información.*
> [For me, Spanish research requires one to look more in the Internet than other sources of information.]

They also recognized the difficulties that smaller cultural groups faced to be heard in modern information societies:

> *Mucha de la gente, los pobres o los que no tienen accesso a educación o tecnología, no puede publicar sobre un tema. Sus historias son importantes pero ellos no tienen voces.*
> [Many people, the poor, or those who don't have access to information or technology can't publish on a topic. Their stories are important but they don't have a voice.]

In this way, it is clear that students are using these class experiences to start to reflect on differences between doing research in English language contexts and the difficulties and realities of research in the Spanish context. Most important, these challenges have made them question and think about their Anglophone privilege and how this plays out in information systems; a step on the road to developing true transcultural competence.

CONCLUSION

In conclusion, this chapter details the development of IL instruction strategies that focus on developing transcultural competence. The internationalization of campuses means that multicultural information realities are more likely than ever before. In turn, this necessitates an approach to IL that takes these contexts into account. This also means that librarians may have to rethink their approach to IL, moving beyond database navigation toward a deeper understanding of students' needs and capacities in today's information societies.

Google, as a core source in many workplace and academic information environments, serves as an ideal basis for thinking about questions of transcultural competence. At the same time, this chapter is not just about Google; instead, we must teach students to critically engage with all information systems. However, Google's dominance means that any bias has a much greater effect than with any other search engine, which implies that it is a good place to begin. Ultimately, however, this chapter draws attention to the tensions between the perceived homogenizing effects of globalization and the underlying diversity that can only be grasped through transcultural education. A careful study of Google can offer students and librarians alike an excellent entry point into this complex subject.

REFERENCES

American Council on Education (ACE). 2012. *Mapping Internationalization on U.S. Campuses.* www.acenet.edu/news-room/Pages/2012-Mapping-Internationalization-on-U-S--Campuses.aspx .

Baker, Paul, and Amanda Potts. 2013. "'Why Do White People Have Thin Lips?' Google and the Perpetuation of Stereotypes via Auto-Complete Search Forms." *Critical Discourse Studies* 10(2): 187–204.

Gillespie, Tarleton. 2013. "The Relevance of Algorithms." In *Media Technologies*, ed. Tarleton Gillespie, Pablo Boczkowski, and Kirsten Foot. Cambridge, MA: MIT Press

Graham, Mark, and Matthew Zook. 2013. "Augmented Realities and Uneven Geographies: Exploring the Geolinguistic Contours of the Web." *Environment and Planning A* 45(1): 77–99.

Green, Lelia. 2014. "The Internet: An Introduction to New Media." www.scribd.com/doc/235144422/The-Internet-an-Introduction-to-New-Media .

Grimmelmann, James. 2013. "Speech Engines." *Minnesota Law Review* 98: 868–952.

Halavais, Alexander. 2013. *Search Engine Society*. Cambridge, UK: Polity.

Hannak, Aniko, Piotr Sapiezynski, Arash Molavi Kakhki, Balachander Krishnamurthy, David Lazer, Alan Mislove, and Christo Wilson. 2013. "Measuring Personalization of Web Search." In *Proceedings of the 22nd International Conference on the World Wide Web*, 527–38. http://dl.acm.org/citation.cfm?id=2487788 .

Hicks, Alison. Forthcoming (a). "Broadening the Landscape: Information Literacy in Foreign Language Education." *NECTFL Review*.

———. Forthcoming (b). "Knowledge Societies: Learning for a Diverse World." In *Not Just Where to Click: Teaching Students How to Think about Information*, edited by Heather Jagman and Troy Swanson. Chicago, IL: ACRL.

Jeanneney, Jean-Noel. 2005. "Quand Google défie l'Europe." *Le Monde*, January 22.

Jiang, Min. 2014. "The Business and Politics of Search Engines: A Comparative Study of Baidu and Google's Search Results of Internet Events in China." *New Media & Society* 16(2): 212–33.

Modern Language Association (MLA). 2007. *Foreign Languages and Higher Education: New Structures for a Changed World*. www.mla.org/pdf/forlang_news_pdf.pdf .

Morozov, Evgeny. 2013. *To Save Everything, Click Here: The Folly of Technological Solutionism*. New York: PublicAffairs.

Purdy, James. 2012. "Why First-Year College Students Select Online Research Resources as Their Favorite." *First Monday* 17(9). http://firstmonday.org/ojs/index.php/fm/article/view/4088/3289 .

Swanson, Troy. 2010. "Information Is Personal: Critical Information Literacy and Personal Epistemology." In *Critical Library Instruction: Theories and Methods*, edited by Maria Accardi, Emily Drabinski, and Alana Kumbier, 265–78. Duluth, MN: Library Juice Press.

Sweeney, Latanya. 2013. "Discrimination in Online Ad Delivery." *Queue* 11(3). http://queue.acm.org/detail.cfm?id=2460278 .

Umoja Noble, Safiya. 2013. "Google Search: Hyper-Visibility as a Means of Rendering Black Women and Girls Invisible." *InVisible Culture* 13. http://ivc.lib.rochester.edu/portfolio/google-search-hyper-visibility-as-a-means-of-rendering-black-women-and-girls-invisible/ .

United States Census Bureau. 2011. *Language Use in the United States*. www.census.gov/hhes/socdemo/language/ .

Chapter Thirteen

Google in Special Collections and Archives

Michael Taylor and Jennifer Mitchell

Google has become a leading force in enhancing access to special collections. Although library-based catalogs can be extremely helpful in locating archival resources, students, scholars, and other researchers are increasingly discovering these types of materials via alternative means. A simple Google search is a quick and, in some cases, wise strategy. From the perspective of special collections staff, Google and its family of applications are also making a big difference. More partner than competitor, Google brings great promise to a corner of the library world that is actively rethinking its mission and service model in order to ensure equality of access and reach a greater audience than ever before. This chapter will offer a brief discussion of the ways Google enhances and improves the archival research process. It will also provide practical suggestions as to how people working in special collections might incorporate Google services into their daily tasks, as well as some food for thought about their possible adverse effects.

HOW GOOGLE HELPS RESEARCHERS

At the beginning of the research process, archival researchers face two challenges. The first is to locate a repository that contains resources pertinent to their topic. Unlike books, archives are one of a kind and often end up in unexpected places. Traditional ways researchers have found material include referring to a printed archival directory in which repositories list their collecting areas; mining the bibliographies of secondary sources to see where other scholars have found materials; asking an expert on the topic for advice; or using an electronic union catalog such as OCLC WorldCat, Archive Grid, or Archive Finder to identify a repository with relevant holdings.

Although these are all sound strategies, more and more researchers are making Google their first stop in the hunt for primary source materials. Since the 1990s, archives have been working to convert their finding aids—descriptive inventories of collections—from paper to electronic versions, utilizing Encoded Archival Description (EAD) or word-processor-generated PDF files, and making them available on their websites. This aids discovery in several ways:

- No matter how small or off the beaten track an archive may be, if it has posted its finding aids to the web, Google can locate them.
- Even many large archives do not report their holdings to archival directories or union catalogs, or do not update the information on a regular basis. Thanks to Google, which constantly crawls the web for new content, archives are better able to keep the world "in the know" about their collections.
- Google searches the keyword-rich full text of finding aids, as opposed to the relatively brief collection summaries found in library catalogs.

The second major problem that archival researchers face is library catalogs that are not well equipped to search for manuscript materials. The catalogs that most libraries use were designed to find books, which can usually be described in a few fields. Archival collections, in contrast, frequently require longer descriptions than are impractical to include in a traditional catalog record. While some institutions have adopted advanced catalogs that are able to pull keywords from the full text of finding aids, others have not yet taken this step. Some have found an alternative in Google's custom search engine. By building one of these into a library's website and programming it to restrict its searches to local finding aids or web pages, you will bypass antiquated catalogs and, at the same time, solve the problem of Google's main search engine returning too many irrelevant hits. Although this is not an ideal solution since it creates one more place for researchers to have to look, it is an alternative that you might want to consider. Also, if you have not yet converted all of your finding aids to a format that Google can search, you may mislead researchers into thinking they are able to search everything in your collection via Google, when, in fact, they may still have to refer to your local catalog for some materials.

Google's ability to search archival resources across and within institutions is not the only way it helps researchers. In addition to making finding aids more accessible, Google has also raised the profile of specialized "in-house" research tools that many people otherwise would not know about. These include subject guides, supplementary inventories, and home-grown databases or reference resources. Novice and even expert researchers, moreover, benefit from the greater flexibility Google offers in keyword searching. Whereas many library catalogs and electronic archival directories rely on controlled vocabularies and specialized terminology, Google allows for the use of more natural language and a broader range of terms. Its autocomplete and search suggestion features improve the chances of relevant results regardless of spelling errors or the inexact usage of terms that could otherwise lead to a dead end. At the same time, Google retains some of the handy features of library catalogs, such as the ability to limit a search by language or to use Boolean operators. (Researchers unfamiliar with how to conduct a Boolean search can utilize Google's Advanced Search, which includes set fields.) While some catalogs are better than others at searching for archival materials, few if any can compete with Google.

HOW GOOGLE HELPS SPECIAL COLLECTIONS LIBRARIANS

In some ways, Google's effect on the daily operations of special collections libraries has been no different than in other types of libraries. The Google search engine has obviously made simple fact-finding more efficient, whether at the reference desk or behind the scenes in areas such as technical services. In other respects, though, special collections staff are tailoring Google and its growing family of applications to fit their own needs.

Processing and Cataloging

Manuscript processors rely on Google to track down information to include in finding aids' biographical and descriptive notes fields. These help researchers understand who created a collection or what its historical significance is. Oftentimes information on obscure individuals is not available in printed resources, or would be challenging to find. Google's ability to tap into a wealth of genealogy websites and other biographical resources has made searching for vital statistics on little-known historical figures much easier. Although searching for a personal name online can bring back a tidal wave of results from social media and news sites, you can use a Boolean operator, such as *-site:facebook.com,* to omit websites unlikely to contain information on historical figures. Also consider starting your search with Google Books and adding an appropriate date range.

Rare book catalogers, similarly, use Google to answer questions related to an individual book's history, such as its author, illustrator, binder, or a previous owner (provenance). Special collections catalogers of all sorts turn to Google to solve a common problem—how to correctly spell names, places, and terms. These have often changed over time, vary in form from language to language, or were simply misspelled in an original document. Google's search suggestion feature can help determine the correct, modern, or standard spelling.

Travel the World with Google

Google Maps has many uses and often points special collections staff in the right direction. Catalogers and reference librarians, for example, frequently encounter materials such as letters, diaries, and photographs that contain vague or obscure place names. A search in Google Maps can help establish a more precise location, especially if other facts are known. The application's street view feature can be used in various ways. For instance, an archivist might need to identify a building in a photograph or determine whether a previous identification is correct. If the building is still standing, Google Maps' street view can provide the answer. Some innovative archivists are using the "My Maps" feature to create customized maps that can be added to finding aids. Are you describing a collection of materials such as architectural records or land surveys? Consider making your finding aid more dynamic by including a map.

While Google Translate often returns imperfect results, it is a tool that people working in special collections sometimes find helpful. You might use it, for example, to establish basic facts, such as what language a book or manuscript is written in, or to translate key words that can help you get the gist of what the material is about. One issue that special collections libraries are facing is the problem of collecting materials that document cultural groups, such as immigrant communities, whose first language is not English, but then providing access in English only. The ideal solution is to offer multilingual discovery tools. As a courtesy and sign of respect, these should be written by a native speaker if at all possible. Although Google Translate is improving and may be able to help, it is still limited in its ability to capture the nuances of a language. In a pinch, it may be a quick solution, but use with caution.

Google and Reference Services

When it comes to reference services, Google's biggest impact, of course, has been its main search box. In special collections, librarians can use it to find information not only outside of their own collections more easily and efficiently, but also within. Patrons can do the same. In some respects this transparency and ease of access may be said to have reduced the need for traditional reference services, but at the same time, it has also resulted in more people finding collections relevant to their research. Requests for basic facts may be in decline, but as you put more information about your holdings into an online environment where a search engine like Google can discover and deliver it to the world, be prepared for more questions from non-local researchers, more requests for access (including photocopies and scans, which can be time-consuming to make), and more applications for permission to reproduce digitized materials.

Library managers can assume that their reference staff will be familiar with Google's basic search. Its advanced features, however, are not yet common knowledge. Consider making them part of the training your employees receive. For example, new reference librarians are typically asked to spend time familiarizing themselves with the reference books found in many special collections reading rooms. Did you know that Google Books is likely to contain many of the same titles, especially biographical dictionaries, local histories, bibliographies, and rare book reference sources? Staff might want to identify the most useful resources in your reference collection and check to see if Google Books has digitized them. You may be surprised at what is available, and the ability to conduct keyword searches is sure to improve the quality of your reference services.

Use Google to Share Your Treasures

From the perspective of special collections outreach and instruction, Google has much to offer. Several of its applications, in particular, are helping repositories of rare books and archives connect to a new generation of users. YouTube, owned by Google since 2006, is now widely used to host short library-related videos. General orientations to special collections facilities are common and can be a great way to help inexperienced researchers feel less intimidated about making their first visit. Do you frequently have patrons asking for a tour of your facility? If so, consider creating a YouTube virtual tour that shows what goes on behind the scenes. Curator interviews, recordings of guest speakers, instructional videos showing how to use the catalog or properly handle rare books and manuscripts—the possibilities are limitless. Once you have uploaded the videos to YouTube, you can add titles, descriptions, and tags to increase their discoverability. The site provides statistics on each video that will help you determine number of views, viewer locations, audience retention, and referrers, among other things. In addition to reaching out to patrons, online videos can also be a great form of donor relations. Administrators, too, may see them as a sign that you are doing your part to engage the public (including online learners) and be a "good citizen" of your institution or community.

Blogs are another way a library can connect to the wider world. Although there are many different blog sites, one of the most popular—Blogger—has been owned by Google since 2003. As with YouTube videos, the uses for blogs as a way of promoting special collections are limited only by your creativity, but some common uses include:

- Posts about new acquisitions and recently processed collections
- "Spotlights" on interesting or unusual materials worthy of greater attention
- Guest posts by researchers eager to share their finds
- News about upcoming events or exhibitions
- Announcements of new digital collections
- Stories about books or articles written using your institution's materials

Patrons can subscribe to the blog's feed, and you can link to blog posts from other social media outlets such as Facebook and Twitter to help keep your users up to date about what's going on in your library.

Special collections resources can also be promoted in a fun, interactive way by using Google Earth, a Geographic Information Systems (GIS) tool. Although its learning curve is higher than other Google applications, the results can be very rewarding. One approach, for example, would be to take a digital image of a historical map and overlay it onto a modern satellite image of the same area. You could then pin scans of letters, photographs, engravings, ephemera, or other materials from your collection onto the map to show how they relate to a geographic area or how that area has changed over time. If desired, you can create an animated tour that moves viewers through the three-dimensional space, either aerially or at street level; audio narration can also be added. To see what others have done and get ideas for your own projects, visit the Google Earth Outreach page.

Measure Your Impact

Other Google applications that reference and public services staff in special collections might consider using include Google Analytics and Google Forms. Are people utilizing the resources you spend so much time putting together? Who is using them? How are they finding you? Do you want to create a survey to get patron feedback on a new or existing service? If assessment is one of your library's goals, Google can help.

To Buy or Not to Buy?

Identifying materials to add to your collection is best done by looking at a printed dealer catalog or specialized websites that dealers use to advertise their stock, such as the Antiquarian Booksellers' Association of America (ABAA), the International League of Antiquarian Booksellers (ILAB), viaLibri, or AbeBooks. Google, at present, does not have a good way to limit your search to sellers' listings. However, when considering a purchase, Google is still something you might want to use for other reasons.

You should always research a book's price before buying it. The sites listed above are better places than Google to do price comparisons of copies that are currently for sale; as for auctions, most people use American Book Prices Current. A simple Google search, however, may be able to turn up records for auctions that took place very recently. You can use these to figure out how much a seller paid for a book or archive and then decide whether his or her price tag is reasonable. Marking up the price of materials is to be expected (after all, dealers have to earn a living, and they are performing a valuable service by finding materials). But don't be duped: if the price has been tripled or just seems suspiciously high, approach with caution.

The idea that projects like Google Books are going to reduce libraries' need to purchase materials (i.e., "just wait for Google to digitize it") is a contentious but widely held belief. When it comes to expensive materials like rare books, it is certainly tempting to refer researchers to open-access digital libraries. This is not necessarily a bad thing, for in addition to being free, they offer full-text searchability, perhaps their biggest advantage. From the perspective of rare books, however, there are several distinct drawbacks to Google Books and other digital libraries:

- A digital copy gives no sense of the book as object. For example, it may not convey a book's size, shape, binding, printing style, paper quality, or other elements of its design that scholars can use to study its history.
- Google Books shows only one particular copy. Although this may be good enough for most users' purposes, textual variants and things like notes, inscriptions, and other signs of reading are sometimes as important to scholars as differences between first and later drafts of a famous author's manuscripts. When it comes to rare books, variety is the spice of life.
- Books are sometimes poorly digitized. Blurry images, defective copies, unopened folding plates, missing plates, photographers' fingers—anyone who has spent a lot of time using Google Books will have noticed these kinds of problems.
- There is no guarantee that the Google Books copy will always be there. As unlikely as it is to go away, the recent economic recession has shown the risks associated with the "too big to fail" mentality.

That said, you might decide that what your patrons really want is full-text searchability. If that's the case, Google Books can help you "get more bang for your buck" by allowing you to spend more of your limited funds on the things you want the most. On the other hand, be sure to take into consideration a book's physical aspects and ask yourself whether that could be of interest to researchers. Paradoxically, examining a digital copy and weighing its pros and cons may provide the justification you need to buy a physical copy. When all is said and done, Google Books is perhaps best thought of as an *auxiliary* to rare book collections rather than a substitute.

Google Can Catch the Bad Guys

One thing that special collections libraries have to worry about that digital libraries do not is security. Hopefully, you will never have to investigate stolen materials, but if you do, Google is a good place to start. Theft has always been a problem for libraries. In the Middle Ages, it was so common that librarians chained books to the shelves to deter readers from walking away with them. Today, we are able to rely on security cameras and high-tech registration systems, but stories about stolen books and manuscripts still make their way into the news on a regular basis. This is probably not, however, because incidents of theft are rising. In part, it is because thieves have a new and powerful enemy—the Internet search engine.

There has never been a worse time to try to sell stolen archival materials on the open market. Search engines like Google cannot prevent the initial theft, but they make it hard for the thief to dispose of materials in a way that does not arouse suspicion. Manuscripts, which are by nature unique, are especially hard to sell online. There have been several cases in recent years of manuscript thefts going unnoticed by archives staff, but then being detected by vigilant scholars who suspected that something was amiss. As a general rule, the more online documentation that exists to tie a manuscript to its rightful owner, the better. This might include:

- *Electronic finding aids*. These are publically available, virtually impossible for an outside thief to alter, and easy for dealers of archival material to locate via a Google search if they suspect someone is trying to sell their stolen documents.
- *Scholarly citations*. If past researchers have used material that has gone missing, they may have noted its proper location in their publications. Google Books and Google Scholar make it easy to search for citations of archival materials in books, theses, journals, and other publications.

In comparison to manuscripts, stolen books are more difficult to trace. Even rare books usually exist in multiple copies, and it can be tricky to determine whether a book listed for sale online is the same copy that is missing from a library's stacks. The first line of defense is full-level rare book cataloging. This may not be practical for every book in your collection, but with your most valuable books, you should try to record as many unique details as you can—inscriptions, markings, bookplates, unusual aspects of its binding, and anything else that might be helpful in identifying it. If a book is ever stolen and advertised for sale online, these details can help you get it back. While a smart thief would not advertise features of a book that could be used to trace it to its rightful owner, Google, which searches online bookselling sites and some auction records, can help catch more careless thieves. It can also be helpful in situations where a book is being sold online by an honest bookseller who acquired it from someone else, unaware that it was stolen property. He or she may then provide a full description, which you can use to establish that the book belongs to your institution.

CONCLUSION

One of the keys to Google's success has been its versatility. It is hard to imagine not being able to find a use for it, no matter who you are or what your business/organization is about. This chapter has suggested a few of the more obvious ways special collections libraries might use it, but there are doubtless many more. Hopefully you will further explore Google's services and come up with ideas of your own.

Chapter Fourteen

Public Library Summer Reading Registration on Google Forms

Deloris J. Foxworth and Roseann H. Polashek

In its infancy, the summer reading program at Scott County Public Library in Georgetown, Kentucky, consisted of a few hundred children (sample enrollment for 1979: 234; 1983: 324; 1990: 533). In 2013, the library distributed 2,465 logs to readers. This significant increase is due in large part to the tremendous growth in the community. However, the inclusion of adults and teens and changes in the registration process may also play a role in the large increase.

In 2011 the new summer reading coordinator revised the enrollment process. Readers registered for summer reading by simply picking up an age-appropriate reading log at the library instead of registering their name, age, and reading goal. This change, while easy to manage, left the library with little usable data. Online registration was introduced in 2014 using Google Forms to obtain better data. In this chapter, the authors provide a history of the Scott County Public Library's summer reading program; establish the need for online registration; explore available options; summarize their decision; explain how to design and use Google Forms; share their results; and offer ideas for using the data.

SOME HISTORY OF SCOTT COUNTY PUBLIC LIBRARY'S SUMMER READING PROGRAM

According to retired library director Earlene Arnett, Scott County Public Library (hereinafter SCPL) began its summer reading program in 1974. In the beginning, only children were included, but today the program has expanded to include babies, preschoolers, tweens, teens, and adults.

Initially children registered with a library staff member, set a reading goal, and upon completion of the goal were honored in the library through a visual display of their accomplishment. As summer reading gained in popularity, Arnett said that SCPL joined a regional coordination effort of Bluegrass North and Bluegrass South 1976 and later joined the statewide summer reading campaigns. The coordinated efforts among the regional and state summer reading campaigns allowed for shared resources with regard to programming and other materials. Individual libraries were responsible for registering and counting the number of readers who participated.

As the number of participants grew, SCPL continued to have readers register and set reading goals. However, the rewards went from recognizing individual readers with library displays to conducting drawings for a small amount of larger prizes, to ultimately hosting a massive pool party for all readers who read over five books during summer. In more recent years, SCPL also introduced a reading competition among the county's elementary schools. To date, the reading competition has been the primary use for data collected during summer reading other than merely keeping track of the number participating for state reporting.

In 2011, since SCPL had very infrequently used the data collected from summer reading, the new summer reading coordinator elected to forgo traditional registration and goal setting for simplification. Instead, the library would count the number of logs distributed and only record names and school information for the purpose of the reading competition.

ESTABLISHING THE NEED FOR ONLINE SUMMER READING REGISTRATION

While numbers for summer reading continued to go up in the county, SCPL, like many other libraries, had been experiencing a decrease in items checked out and in visits to the library. This got the summer reading coordinator thinking about how the success of summer reading could translate into more success throughout the year. The answer took the summer reading coordinator back to summer reading registrations, so the library could collect patron contact information, such as e-mail addresses, for target-marketing of library programs and performers. SCPL could also use the collected information to keep patrons aware of the upcoming library expansion, send out library e-newsletters, and inform of any possible program changes or library closures due to construction.

INVESTIGATING OPTIONS

In 2012, the summer reading coordinator discussed with the newly hired technology manager about the possibility of reviving summer reading registration at SCPL and making it digital. The summer reading coordinator knew there was the potential to get even more sign ups if registration went digital because patrons would no longer have to come inside the library to register. The online registration would also allow the library to collect more information about participants and their programming preferences.

The technology manager began researching options. Many libraries had successful digital summer reading sign-ups, but how did they do it? What information did they collect and what did they do with the information?

The technology manager scoured the Internet looking for examples from other libraries. Forms seemed to vary in what information they collected and how they were designed. Following the technology manager's initial research she discovered three main options for online registration: building a form from scratch, using a reservation system, and using Google Forms.

Building a Form from Scratch

The first option was to build a form from scratch on the library's website to collect data. This option quickly died as the library staff was not familiar enough with Dreamweaver and coding languages to create a form from scratch. However, the technology manager found a nearby library that was willing to share a form they created using PHP, a server scripting language used in dynamic and interactive web design (W3Schools 2014).

The PHP file was very complex. It was tied to the other library's Integrated Library System (ILS) and authenticated patrons' library cards when registering. It did not collect contact information from the patron since it was going through their internal system.

Upon review and consultation with the summer reading coordinator, the technology manager believed this form to be incompatible with SCPL's goals. While technically it would work with SCPL's website and even ILS, after some tweaking, it did not collect the appropriate bits of information SCPL was seeking. The form also seemed too complex for the staff to troubleshoot.

This put the technology manager back at square one, leading to the second option, a web-based solution that could be accessed from the library's website. The technology manager explored two reservation systems: Evanced Solutions by Demco and Constant Contact's EventSpot.

Using a Reservation System

The technology manager's investigation of some reservation systems left her unsatisfied. While these products appeared user friendly for both the creator and the end user, they collected the wrong types of data and came with a significant price tag. For example, Evanced Solutions by Demco has a product specifically for summer reading that allows readers to "track their reading, monitor the prizes they've won, write reviews, and much more" (New York State Education Department 2014). Since SCPL does not store patrons' circulation history for privacy and legal reasons, the technology manager felt this product was incompatible with the library's privacy policy.

Constant Contact's product, EventSpot, allows businesses to create invitations and registration forms, process payments, and much more (Constant Contact 2014). However, the library was not collecting money, so this product seemed excessive based on the library's needs.

Using Google Forms

Still not satisfied that she had found a product that was user friendly and customizable to meet the library's simple needs, the technology manager found another option. The technology manager had a personal Google account and had used Google Docs periodically over the years, so after learning some other libraries had successfully used coordinating product Google Forms for registration purposes, the technology manager was considering the option. However, since the form and the data collected are stored online, security and privacy of the data was a large concern.

Upon investigation, the technology manager learned that access to Google Forms and the data generated can be controlled by the form owner. There are three different access options: *public access*, *anyone-with-link access*, or *specific invite access*. *Public access* means the file can be located through an Internet search. *Anyone-with-link* means anyone with the exact link address can access the file. *Specific invite access* means anyone the owner has invited to view the file can access it, but it is otherwise secure ("Change Your Sharing Settings" 2014). Google also claims that data is stored in "geographically distributed data centers" with resiliency and redundancy measures in place to prevent loss due to equipment failures and environmental risks. Third-party audits verify Google meets SSAE 16 and ISAE 3402 requirements for protecting data ("Your Security and Privacy" 2014).

Google Forms, part of Google's SkyDrive product, is free and easy to set up and use to collect data. In the pages that follow we will guide you through setting up your own Google Form.

ACCESSING GOOGLE FORMS

Library staff must have a Google account to create a form. Libraries using Google's Gmail service should already have access to Google's online services, but those without e-mail can create an account for free. Creating a Google account gives individual access to Google Drive as well as many other free Google features. Google Drive is a web-accessible storage solution to store, manage, and access files from any device. Google Drive gives users' access to Google Docs and to Sheets and Slides, where users can create, edit, and share documents, presentations, spreadsheets, forms, and drawings ("Google Drive Overview" 2014).

For libraries, without access to Google, visit accounts.google.com and click on Create an Account toward the bottom of the screen. Remember, the account needs to be set up for an individual (each library staff member could create an account for free access to e-mail and other features). Once on the Create Your Google Account screen, enter the necessary information, including a username (which will then be your Google e-mail address). If you do not want another e-mail address, simply click "I prefer to use my current e-mail address" on the form under Choose Your Username. The form then replaces the username field with a field to enter your existing e-mail address.

Create a Form

Go to Google.com. Click on Sign In. Once logged into a Google account, select Drive from the Google menu on the upper right side of the screen. Next, click on Create. From the drop-down menu select Form. A pop-up window will appear. Inside the pop-up window there is a place to provide a title for the form and to select a pre-designed form or create a default form. After clicking OK, Google will prompt you to enter your first question.

To choose any question type, select it from the drop-down menu. There are nine types of questions to choose from:

1. Text questions: Respondents can provide short answers ("Add and Edit" 2014). Text questions can be used to collect small, unique identifiers such as first and last name.
2. Paragraph text questions: Respondents can provide longer answers ("Add and Edit" 2014) so these questions can be used to collect opinions or feedback.
3. Multiple choice questions: Respondents choose one answer from among provided options ("Add and Edit" 2014). Multiple choice questions can be used to limit the answers received. For example, if the library had promoted an event through certain channels, you could use multiple choice to limit responses.

4. Checkboxes: Respondents can check all options that apply ("Add and Edit" 2014). Checkboxes can be used to ask how readers learned about a program or whether they would like more of certain types of programs in the future.
5. Choose from a list: Respondents are given a drop-down list of options to choose from ("Add and Edit" 2014). This question can be used to collect general age information or grade information about participants. Google uses a drop-down list for question type.
6. Scale: Respondents can rank their response on a scale of numbers ("Add and Edit" 2014). A scale question can be used to get respondents' overall impression to a summer reading kick-off program.
7. Grid: Respondents select a point from a two-dimensional grid ("Add and Edit" 2014). Grids can be used to create Likert-scale questions to determine how satisfied or unsatisfied respondents are with summer reading events.
8. Date: Respondents pick a date from a calendar ("Add and Edit" 2014). Date could be used for readers to record when they complete books.
9. Time: Respondents can select a time of day or duration of time ("Add and Edit" 2014). This could be used for readers to record the amount of time spent reading.

Each question, no matter the type, can be made a required response by checking Required Question. If required, respondents will be prompted to answer before Google will accept the form. Once you have entered the necessary information, click Done to add the question to the form.

To add another question to the Google form, click Add Item or click the drop-down menu button next to Add Item to be prompted to add another question.

By clicking on the drop-down button, the range of question types will be displayed. Select the type of question you want to create. Fill out all the necessary information for that question type and then click Done to add the question to the Google form. Don't forget to check Required Question if you want to make the question mandatory for submission.

Cleverly, Google Drive automatically saves files so you do not have to save regularly.

Designating Responses

Before you can review responses, you must designate how and where responses should be collected. Click on Choose Response Destination, and then choose to Create a New Spreadsheet or Create a Sheet in an Existing Spreadsheet. Another option is to Keep Responses Only in Forms. Keeping responses online in the form saves a summary of the responses on the form, but to view individual results, each individual record must be downloaded. ("Choose a Form Response Destination" 2014)

Creating a new spreadsheet or sheet in an existing spreadsheet saves each response as a record in a spreadsheet that is readily available on Google Drive. Summary data is not provided, until the data collector sorts or calculates data in the spreadsheet ("Choose a Form Response Destination" 2014). For example, the summer reading coordinator sorted the data by grade to determine the number of participants in different groups (babies, preschoolers, tweens, teens, adults). She did this by selecting the grade column and clicking the A/Z sorting button on the menu. Once sorted, she could then use the sum button on the menu to count the number in each group. To count, go to an empty cell, click the drop-down menu next to the Sum button and choose Count. Next, select all the cells you want to count or enter the range of cells in the formula cell ("Add Formulas to a Spreadsheet" 2014). Press Enter to get total number in that group. Repeat for each group.

Accepting responses, or making the form active for use, can be turned on or off with the simple click of the Accepting Responses button. You may elect to turn off responses if the window for the survey has not yet opened. Google allows you to create a message that will be displayed if users attempt to access the form when it is not currently accepting responses.

Adding the Form to the Library's Website

There are two ways to add the form to the library's website. One simple way (and the method the SCPL used) is to click on Send Form found in the upper right side of the open form, or under File, and then Share Link to add the link to the library's website. SCPL's web staff did this by putting a hotspot on a graphic on the home page. This could also be done as a text link.

The Google form can also be embedded into a web page. Click Embed from the Send Form pop-up window. Set the size for the form to display. Then copy the HTML code from the share window. Paste this code into the HTML code of the web page where you want the form to display. The form should look as though it is a part of the library's website.

Inviting Others to View or Download Data

An added benefit of the online storage of Google Forms (non-machine-specific storage) is ease of sharing the form and results with library staff. The creator/owner of the form and spreadsheet can grant other staff members viewing and/or editing access to the files. To share files, the creator/owner should log in to their Google account and go to My Drive. While in My Drive, you will see a list of all files saved to the Drive. Check the box next to the files you want to share. When one or more files have been selected, you will see a graphic menu of options appear above the list. To share the file(s) with others, click the icon with the person and the plus sign.

After clicking the sharing icon, a Sharing Settings dialog box will appear. At the top it has the link to share with collaborators. Next is a section that lists who has access to the files. It should list you as the owner. To add new people to this, enter the individual's name or e-mail address in the box under Invite People. Once you start typing a name or e-mail address, next to the box you will see a drop-down menu to specify what kind of access you want the person to have. Right now you can grant the person access to edit. Once they are added you can change that person to become owner if you want, but only one person can be owner at any given time.

The share feature should only be used to share with staff and other persons needing it for administrative functions. These files should not be shared with patrons. Patrons should be given the link to access the Google form only and should not be given access to the spreadsheet. Google transfers their information to the spreadsheet using the Google form.

EVALUATING THE USE/SUCCESS/ADOPTION OF GOOGLE FORMS

The initial use of Google Forms to register readers in the summer reading program seemed successful. As of July 17, 2014, the summer reading coordinator reported 2,032 participants, 1,560 of which were online registrants. The online registration number includes adults, teens, and children who registered online. Adult patrons were not required to register online, however, 200 did, and an additional 272 picked logs up at the circulation desk. Children and teens were supposed to register to receive coupon packets and reading logs and to participate in the reading race for the School Summer Reading Trophy.

Participation appears to be down in 2014, as of mid-July. Only 2,032 logs were distributed in 2014, compared to 2,465 logs in 2013 for ages eighteen and under. Fewer summer camps and fewer daycares participated in the SCPL reading program in 2014, for unknown reasons, and is the probable cause for this decline in numbers. The final totals, however, will not be available until summer's end. We will not speculate that the decrease in summer reading participation is due to a move to digital sign-ups, as no one was "required" to sign up online. Library employees completed forms for patrons who were not comfortable with technology, and adult patrons visiting the circulation desk were not even required to sign up. Weather and economic considerations beyond the library's control may also have had an impact.

The Summer Reading kick-off day seemed to go rather smoothly. There were three iPads available for registration. Staff members and teen volunteers were available to help patrons navigate the new system. The sole complaint was from a few patrons registering multiple readers, such as their five children, all at once; a form was filled out for each child. Next year, maybe there should be a field to include multiple young readers in one family. Following kick-off day, patrons registering at the information desk continued to use an iPad to complete a Form for each reader. Most kids were excited to use this technology!

Staff had mixed, but mostly positive, reviews about online summer reading registration. Some staff members thought moving to online registration was a much needed move. Other staff felt it was labor intensive. If patrons were not comfortable using the iPad the staff member would enter the information and then read the information back to the patron before submitting. Staff with limited experience on iPads felt uncomfortable filling out forms or trying to guide patrons through the process. Next year, paper forms for patrons and staff uncomfortable with technology may be introduced. Then a designated staff member could enter information from the paper slips into the Google responses spreadsheet later.

FUTURE USE OF DATA COLLECTED

The purpose of moving to an online summer reading registration was to collect usable information about program participation. The spreadsheet is sorted by date and time, but it can be sorted many other ways to gain further insight into the summer reading population. For example, the timestamp for each entry may reveal valuable information about the best time to hold library programs. Combine the timestamp with the grade category on the spreadsheet and the library has an age range for the audience to target during that time.

The data can be downloaded as an Excel spreadsheet for further use. For example, the e-mail addresses can be checked against the e-mail addresses in the ILS system. The entire list can be imported into Constant Contact, allowing the library to send out e-mails reminding people to turn in their logs, hoping to increase return rates. Next year's summer reading and other library events can be promoted through the information gained from this form, using both the e-mail list and the school list.

The Google form also collected information about how patrons learned about summer reading. That information can be used to determine the most effective forms of promotion for library events. Combine that with the question about what kinds of programming people are interested in and the library can tailor marketing messages and mediums based on the intended audience.

Of course, any mass e-mails would provide the patron with the choice to *opt-out* of future messages!

FUTURE USE OF GOOGLE FORMS

Summer reading registration is just one application of Google Forms. The library has no definite plans to use Google Forms for other projects as of now, but here is a list of some additional uses for Google Forms in public libraries. This list is not exhaustive, so please feel free to devise your own plan for Google Forms.

1. Surveys about programs or collection. The form could be linked to the website or e-mailed to existing patron e-mail list.
2. Registration for programs/events.
3. Drawings/give-a-ways.
4. Collecting topics for future workshops/book clubs.
5. *Contact us* link on website.

SUMMARY

While the Scott County Public Library may not have gotten the significant increase in summer reading registration it had hoped for with the implementation of Google Forms, it definitely has a new tool in its arsenal for collecting and distributing information. The authors feel that Google Forms is a simple, yet effective way to encourage online interaction with patrons, collect data for marketing and collection development purposes, and share information among library staff.

REFERENCES

"Add and Edit Questions, Headers, Images, Videos, and Page Breaks." 2014. *Google*. Accessed July 27, 2014. https://support.google.com/docs/answer/2839737?hl=en .

"Add Formulas to a Spreadsheet." 2014. *Google*. Accessed August 5, 2014. https://support.google.com/docs/answer/46977?hl=en .

"Change Your Sharing Settings." 2014. *Google*. Accessed July 28, 2014. https://support.google.com/drive/answer/2494886?p=visibility_options&hl=en&rd=1 .

"Choose a Form Response Destination." 2014. *Google*. Accessed June 28, 2014. https://support.google.com/docs/answer/2917686?p=forms_response&rd=1 .

Constant Contact. 2014. "See How EventSpot Works." Accessed July 27, 2014. http://www.constantcontact.com/eventspot2 .

"Google Drive Overview." 2014. *Google*. Accessed, July 27, 2014. https://support.google.com/a/answer/2490026?hl=en .

New York State Education Department. 2014. "Summer Reader Online Registration Tool." *Summer Reading at New York Libraries*. http://www.nysl.nysed.gov/libdev/summer/smreader.htm .

W3Schools. 2014. "PHP 5 Tutorial." Accessed August 5, 2014. http://www.w3schools.com/php/default.asp .

"Your Security and Privacy." 2014. *Google*. Accessed July 3, 2014. https://support.google.com/a/answer/60762?hl=en .

Chapter Fifteen

Seeing Libraries through Google Glasses

Barbara J. Hampton

By the time you read this, smart wearable technology and augmented reality will no longer be limited to the realm of geeks and elites. The Internet-connected world will be accessing information and integrating it into daily work, play, and learning. Library programs, policies, and presentation of information must evolve to incorporate the new devices and to educate staff and patrons alike about their use; many will find these devices on their personal wish lists, too.

Google Glass incorporates features widely available in various other devices: Internet searching, camera, video recorder, text messaging, cell phone calls, video calls, social media, apps, GPS, Bluetooth, and voice recognition. What makes Glass exceptional is the integration of the tools with each other and with hands-free operation, using voice commands, head-tilts, and winks in addition to simple finger taps and sweeps on Glass's frame. It's not intended to replace our current primary tools for text creation, access, and reading. Users wear the transparent Glass screen just above the line of sight, viewed like a twenty-five-inch high-definition display seen from eight feet away. Voice commands and audio from the device are heard via bone conduction, leaving the user's natural hearing unobstructed. Both Glass and the user are "connected" to their environment, their companions, and events around them, something that's been lost with the thumb-typist, hunched over a hand-held device, checking for messages while walking into water fountains (Starner 2013).

WHAT'S THE BUZZ?

The traditional computer "search engine" is becoming a "do engine." Instead of returning a long list of keyworded Internet links, Glass filters computer responses based on the user's situation and goals. This is made possible by the integration of information from many sources, not just Internet pages. Rain is forecast for this evening: Would you like to message the guests you've invited for a barbecue to meet instead at a nearby steakhouse? The holidays are approaching: How about the locations of some toy stores near your office that have just what your little ones are hoping to receive as gifts? Your flight is canceled due to weather problems on the other coast: Do you need a hotel reservation? Sophisticated algorithms allow these next generation devices to anticipate your needs, give you timely notifications, and suggest some options. As a reference librarian said in the 1955 Broadway comedy (later a movie) *The Desk Set* (Marchant 1983, 29):

> Richard [computer engineer]: It does give you the feeling that maybe—just maybe—that people are a little bit outmoded.
>
> Bunny [reference librarian]: Yes, I wouldn't be a bit surprised if they stopped making them.

Not surprisingly, other head-mounted technology is coming to market; some devices are reverse-engineered attempts to copy Glass's features. Others represent years of independent development with quite different functions, such as Oculus Rift, goggles that block all views of one's surroundings and substitute screens with computer-generated images to create a virtual reality environment so complete as to trick the brain, used principally for gaming (Parkin 2014; Rubin 2014).

Glass's tools represent a technological leap generating a new vocabulary and bringing esoteric computer science terms into the vernacular. A brief glossary of some useful terms you'll be hearing and seeing is provided at the end of this chapter.

HOW CAN GLASS BE USED?

Later sections of this chapter paint a detailed picture of Google Glass, its distinctive features and methods of operation, and the applications that have and will be developed for it. Its lightweight system, operated heads up and hands-free, is activated with the spoken hotword "OK, Glass." Users particularly value these features where hands and eyes are needed for priority tasks. The camera and screen views are very stable, thanks to the human body's internal balance mechanisms for the head. Most users quickly find themselves visualizing ways Google Glass can improve their personal and work life. If you are intrigued (or just curious), I urge you to meet with a Glass Explorer, to try out Glass for yourself. If you don't know a Glass Explorer personally, watch for events at area colleges and libraries, professional workshops, or conferences. Also, check out relevant community groups on Google+, such as:

- Google Glass in Education
- Libraries and Glass
- Healthcare Glass Explorers
- Google Glass Surgeons
- Google Glass Talk
- Wearable Technology
- Google Glass Explorers and EMS

and blogs that have good coverage of Google Glass and your particular interests:

- TechCrunch
- Gizmodo
- Wired
- MIT Technology Review
- AmericanLibraries
- Public Libraries Online
- Chronicle of Higher Education Wired Campus
- eCampus News

HOW CAN LIBRARIANS USE GLASS?

In the twenty-first century, library users recognize librarians as resources for information and technology solutions, not just guardians of the books and the reading room. By connecting with new developments in technology, we not only provide better services to our fans, we market the library to some who may have thought that libraries were extinct in the age of the Internet. Those who have no plans to own Glass may still have questions about the device. If you've got Glass, wear it loud and proud.

You and your shelvers will be pleased to see that one of the currently available apps, Shelvar, is specifically designed for libraries. By scanning a shelf of items tagged with QR codes on the spine, Shelvar can spot any one that is out of sequence and show the shelver the correct location for the book. Imagine retrieving all those unfindable books that have been mis-shelved! Uploading a library floor plan and coding the shelves can create accurate way-finding for users searching complex open stacks (Bruno 2014; Shatte, Holdsworth, and Lee 2014), as well as the basic conveniences that patrons need: restrooms, copiers, coffee, and so on.

A number of apps translate text or speech (e.g., Word Lens, Lingua App, Unispeech); this could make the library much easier to use for someone whose first language is not English. And while libraries don't often have a sign language interpreter available, even for group presentations, an app (Watch Me Talk) has been developed that generates real-time text captioning of speech.

For library instruction, Google Glass's first-person view of a field trip can bring your audience up close (virtually) with libraries and library resources. A visiting rare book display at a museum? An archive or special collection too far for the students? Inspiration from a higher-level library where they may study in the future? A closed collection might admit one librarian more readily than a student group. In addition to a hands-free video recording or individual image, your virtual tour can be streamed live to your class, and you can relay students' questions to an expert on site. A virtual field trip could showcase your own library (Haefele 2014).

WHAT RESOURCES CAN LIBRARIES OFFER CURRENT AND FUTURE GLASS USERS?

The online Google Glass Help Center (https://support.google.com/glass/) has updated information about the features and operation of Glass.

- At the end of a visit to the park, a parent might search for nearby libraries to get information about the KanJam game he saw played there or so his kids could pick out a book for summer reading. Is your library website engaging and easy to navigate on Glass?
- Are publishers offering new book formats that include Glass-friendly links to supplemental material? If you'd like to see Tolkien's map of Middle Earth while reading *The Lord of the Rings*, a Glass user could glance up at the image without leaving the text page.
- Do you inspire users to access information with Glass about the places, events, wildlife, etc., they see on their travels?

As of this writing, Amazon.com offers over two hundred text items referencing "Google Glass" (although quite a few of those are science fiction or opinion essays). Sadly, only one is at an elementary school level (Ventura 2014), and there are few magazine articles for younger readers as yet, either. More sophisticated middle school to high school readers will enjoy Bran-

Figure 15.1. Google Glass vignette of 1841 wooden whaling ship *Charles W. Morgan*, restored and relaunched in 2014, docking at Fort Adams, Newport, R.I., during the 38th Voyage celebration with inset Google search text result.

wyn's article on DIY Glass (2014). Hartman's forthcoming book on wearable technology will inspire the designers in your library.

Tech-savvy patrons are hungry for organized information about the promise and reality of augmented reality and wearable computers. A fairly comprehensive *For Dummies* book (Butow and Stepisnik 2013) gives detailed instruction on operating Glass. Those interested in writing app programs for Glass using the HTTP-based program from Google, the Mirror API, will appreciate Redmond's guide to the Mirror API (2014). More experienced software writers may use the Android-based Glass Development Kit available online from Google. Those wanting to test out a concept might find the Sim app helpful for "pretotyping."

Many more titles are in the publishing pipeline, some available as Kindle eBooks (denoted with [K] following their entry in the References list below). As you can see from the References list, many news articles about Google Glass and wearable computers are online publications. Help your readers keep up with the latest developments with a Google Glass LibGuide or other pathfinder.

Academic and special libraries are incorporating Glass (Hernandez 2014). Your collection and resource guide should include bibliographies of relevant research publications, information about online communities discussing the Google Glass in their field, and subject-specific research and demonstration projects using Glass. Many can be viewed on YouTube. Andrew Vanden

Heuvel (Google Glass 2013b) took his class to the CERN accelerator in Switzerland. Coaches, athletic trainers, and sports medicine practitioners will be interested in Bethanie Mattek-Sands's (Google 2013c) professional tennis video or the sport and fitness apps (GolfSight, GlassFit, Strava). Health sciences professionals are using Glass with simulation training, documenting and coordinating patient care, and quick reference resources in developing treatment protocols (Albrecht, von Jan, Kuebler et al. 2014; Levine 2014). Glass has helped deaf readers (Jones, Bench and Ferons 2014) and the visually impaired (Yus, Pappachan, Das et al. 2014) and Parkinson's patients (Mohoney 2014).

First responders welcome the mobility and hands-free access with which Glass transmits vital information and documents specific to an emergency situation, without losing sight of the scene around them (Google Glass 2014a):

- Where and in what condition was the patient found?
- What does crash damage tell about possible hidden injuries?
- Where is the nearest fire hydrant?
- Is there a disabled person living in this building?
- Does the person I see match the picture of the suspect?
- What is the layout of this building?

The ability to inspect and document construction and machinery issues also attracts engineers, operators, and contractors to Glass's heads-up, hands-free features (Lee 2014).

THE TECHNOLOGY SANDBOX

Many academic and public libraries offer demonstrations, classes, trial use, and even borrowing of library-owned technology such as eBook readers, tablets, and such. Glass offers similar practical issues, but quite a few libraries have acquired Glass as a library resource (Asgarian 2013; Booth and Brecher 2014; Biemiller 2014).

WHO'S GOT GLASS? WHO WANTS TO GET IT?

Google's "Project X" introduced to the public in April 2012, with the initial "XE" units sold beginning about a year later to those selected (after written application) as Glass Explorers (Google 2012; *Glass Almanac* 2014). I applied to become a Glass Explorer. While at ALA, I tried on Glass, without the benefit of my usual prescription lenses. Fortunately, at that time, Google was rolling out eyeglass frames designed to accept both prescription lenses and the Glass hardware, as well as a list of opticians trained in fitting prescription Google Glass units.

The software for Google Glass receives regular updates, Version 18 being issued in June 2014. Also in June, Google upgraded the hardware for the XE units, doubling RAM to 2GB. Glass accesses its apps through your cell phone (Android or iOS) as well as a mobile hotspot when not on a WiFi network. Hundreds of specialized apps have been released to date, ranging from familiar social networking and newsfeed programs to specialized apps for medical, law enforcement, and disabled services users. Apps are available on multiple sites (and free):

- Google Glassware: https://glass.google.com/glassware
- GlassAppz: http://glassappz.com
- Glass Apps Source: http://www.glassappsource.com
- Glass-Apps: http://glass-apps.org

The exact number of Glass Explorers has not been made public, but it is thought to be about ten thousand. The price has remained at $1,500 for Explorers. Early adopters have included librarians, educators, software developers, doctors, engineers, researchers, artists, entrepreneurs, and, well, explorers! Rumors have suggested that a new hardware version might be released in the next year, and perhaps a price reduction.

WHAT POLICIES ARE NEEDED IN THE AGE OF WEARABLE TECHNOLOGY?

Will there be Glass users needing guidance, supervision, and (gentle) enforcement of civil behavior standards in the library? If you've dealt with users who speak into their cell phones as if using tin cans and string, you know the answer.

Policies should parallel other conduct rules: The library is a public space. Don't interfere with others' use and enjoyment of the library. Do you permit offensive language or name-calling? Do you prohibit carrying or using digital cameras? Librarians take pride in their discretion, but if a local celebrity visits public spaces, he or she will surely be observed by the public. Duplicating copyrighted works is, of course, illegal and unethical: How do you currently enforce this at the copy machine or the computer? (Google Glass 2014b; Sawyers 2013; Shteyngart 2013; Scoble and Israel 2014; "Glass Explorers Do's and Don'ts" 2014; Yus, Pappachan, Das et al. 2014).

Give separate consideration to acceptable behavior for staff, both dealing with the public and in staff-only settings. Some organizations prohibit camera devices where there is access to confidential customer or staff information (e.g., name and address, telephone number, e-mail address, financial transactions, salaries, personnel files, borrowing records). In private meetings, others may suspect that Glass would be used to record audio or images (even when cell phones are also present). Some believe that Glass is a dangerous distraction from driving. Until Glass and other wearable technology become ubiquitous, users should be sensitive to the social distraction it may cause when conversing with others.

If patrons connect Glass to the Internet using the library's WiFi, security risks to the library's IT systems are also a concern. If you have blocks or policies that limit access to some Internet sites, will they be effective for searches done from Glass? Could this be a conduit for malware or data theft? Set these policies in line with other personal devices such as laptops, tablets, and cell phones that have access to the WiFi system.

Use these as guides for your "Glass ceiling."

GLASS GLOSSARY

API: "application programming interface"—specifies how software components interact with each other.

APK: "application package"—installation file for application.

augmented reality: digitally enhanced view of the real world, enhanced with sensory input such as sound, video, graphics, or text layers of digital information, providing context-relevant additional content such as navigation, news, and background information.

bone conduction: transmitting sound with vibration through the user's facial bones rather than through the air, eliminating the need for a microphone for speech or an earphone for hearing.

context aware services: technology services provide more relevant services and content to the user by incorporating user-specific information (location, traffic, weather, date/time, transportation, and information retrieved from devices and uniquely identifiable objects).

do engine: replaces "search engine" results (lists of links) with artificial intelligence–based responses to user's voice input to suggest and/or initiate useful responses.

Explorer Edition (XE): the first public models of Google Glass with delivery beginning in early 2013.

explorers: individuals authorized by Google to purchase, use, provide feedback, and develop new features for Google Glass.

eyeing-up: shifting one's gaze upward to view a head-mounted display slightly above one's direct line of sight.

haptic technology: nonverbal communication using the sense of touch, such as vibration, motions, or tactile forces transmitted by a user or a technology.

heads-up [head-mounted (HM)] display: electronic devices delivering visual results to screens at eye level, either opaque (virtual reality substituting for natural vision), semi-transparent (augmenting natural vision), or through a projection directly onto the retina.

hotword: a word or phrase that triggers a computer action, such as Google Glass's "OK, Glass" activating the basic command options.

immersive virtual reality: computer generated imagery (CGI) displayed on an opaque screen on goggles, blocking views of the user's actual environment and substituting an interactive, three-dimensional digital environment that responds in real time to the user's actions and motions, which can "trick the brain," such as "Oculus Rift."

Internet of Things (IoT): connecting, controlling, and gathering data from devices (computers, smartphones, electronic controls, sensors, etc.), uniquely identifiable objects (GPS locations, RFID, QR code tagged objects, etc.), without human input.

pretotyping: a partially mocked-up version of the intended product or service that can be built rapidly, used to test and collect feedback and usage data without exhausting start-up resources.

side-loading: accessing (and manually installing) an application package outside the Android Market via camera-read QR code, a USB connection, the Windows APK Installer program, etc.

smart glasses: eyeglasses or eyeglass frames incorporating miniature computers and displays, such as Google Glass and its various imitators.

text-to-speech: describes software that "reads aloud" text to audio format.

vignettes: images captured of a Google Glass screen view showing the current view enhanced with a small image of a previous screen view (such as a map, a web image, or text, etc.) in the upper right corner of the image. Vignettes can define a scene or illustrate images, information, and social networking available while users maintain visual connection with people, environment, and activity.

voice recognition: translation via software of a user's spoken words or sounds from the environment into text or computer commands.

wearable computing: miniature electronic multifunction devices capable of complex computational functions, connecting with other computer systems via wireless networks.

wearable technology (aka "tech togs," "fashion technology"): clothing and accessories incorporating advanced electronic technologies, such as smart watches or fitness bracelets, typically developed for specialized purposes such as calculating, vital sign measurements, activity tracking, location tracking, etc.

REFERENCES AND FURTHER READING

Albrecht, Urs-Vito, Ute von Jan, Joachim Kuebler, et al. 2014. "Google Glass for Documentation of Medical Findings: Evaluation in Forensic Medicine." *Journal of Medical Internet Research* 16(2): e53.

Asgarian, Roxanna. 2013. "Arapahoe Library District Invests in Google Glass." *Library Journal* (blog), November 26.

Beilinson, Jerry. 2014. "Total Immersion." *Popular Science* 191(6): 76–81, 118–20.

Biemiller, Lawrence. 2014. "QuickWire: Yale U. Libraries Add Google Glass to Offerings." *Chronicle of Higher Education Wired Campus* (blog), February 6.

Booth, Char, and Dani Brecher. 2014. "Ok, Library: Implications and Opportunities for Google Glass." *College and Research Library News* 75(5): 234–39.

Branwyn, Gareth. 2014. "The (Google) Glass Menagerie." *Make:* 38 (April/May): 34–37.

Bruno, Tom. 2014. "Finding Umberto Eco in the SML Stacks #ThroughGlass." YouTube video, January 27, www.youtube.com/watch?v=4aJ19E8pjKM .

Butow, Eric, and Robert Stepisnik. 2014. *Google Glass for Dummies.* Hoboken, NJ: Wiley. [K]

Cole, Samuel. 2014. *The Google Glass Revolution: How Google Glasses Work and What They Mean for the Future*. St. Petersburg, FL: BMS Publishing. [K]

Denning, Tamara, Zakariya Dehlawi, and Tadayoshi Kohno. 2014. "In Situ with Bystanders of Augmented Reality Glasses: Perspectives on Recording and Privacy-Mediating Technologies." *Proceedings of CHI 2014*, Toronto, ON, 2377–86.

Dublon, Gershon, and Joseph A. Paradiso. 2014. "Extra Sensory Perception: How a World Filled with Sensors Will Change the Way We See, Hear, Think, and Live." *Scientific American* 311(1): 38–41.

Durant, Brad. 2014. *Google Glass: The Ultimate Guide for Understanding Google Glass and What You Need to Know*. Amazon Digital Services. [K]

Glass Almanac. 2014. http://glassalmanac.com/ .

"Glass Explorers Do's and Don'ts." 2014. *Google+ Communities, Glass Explorers*. https://sites.google.com/site/glasscomms/glass-explorers .

Glenn, Brendon. 2014. "What Can Google Glass Do for Medicine?" *Medical Economics*, March 13.

Google. 2012. "Project Glass: One Day Google . . ." YouTube video, 2:30, April 4, www.youtube.com/watch?v=Vb2uojqKvFM .

Google Developers. 2014. *Design for Glass*. https://developers.google.com/glass/design/index .

Google Glass. 2013a. "How It Feels [through Google Glass]." YouTube video, 2:15, February 20, www.youtube.com/watch?v=v1uyQZNg2vE .

———. 2013b. "Explorer Story: Andrew Vanden Heuvel [through Google Glass]." YouTube video, 2:14, May 3, www.youtube.com/watch?v=yRrdeFh5-io .

———. 2013c. "Explorer Story: Bethanie Mattek-Sands [through Google Glass]." YouTube video, 2:35, June 24, www.youtube.com/watch?v=_iMA-y_4KwA .

———. 2014a. "Explorer Story: Patrick Jackson [through Google Glass]." YouTube video, 1:28, January 20, www.youtube.com/watch?v=QPbZy2wrTGk .

———. 2014b. "The Top 10 Google Glass Myths." *+GoogleGlass* (blog), March 20.

Google Glass Books. 2014. *Google Glass: What Is It and How Can It Change Our Lives*. Seattle: Amazon Digital Services. [K]

Haefele, Chad. 2014. "My Week with Google Glass: Library-Centric Thoughts." *Hidden Peanuts* (blog), March 12. www.hiddenpeanuts.com/ .

Hartman, Kate. 2014. *Make: Wearable Electronics: Design, Prototype, and Wear Your Own Interactive Garments*. Sebastapol, CA: Maker Media.

Hernandez, Ernesto. 2014. "Google Glass in the Academic Library." *Journal of Creative Library Practice*, March 25.

Jones, Michael, Nathan Bench, and Sean Ferons. 2014. "Vocabulary Acquisition for Deaf Readers Using Augmented Technology." Paper presented at 2014 2nd Workshop on Virtual and Augmented Assistive Technology (VAAT), Minneapolis, MN. doi: 10.1109/VAAT.2014.6799461.

Lee, Coline T. Son. 2014. "Hands-Free Performance Support." *T+D* 68(4): 108–9.

Levine, Brian A. 2014. "An Ob/Gyn's Journey with Google Glass: Chapter 1." *Contemporary OB/GYN* (blog), January 31.

Levine, Jenny. 2014. "Between Google and a Hard Place." *American Libraries* (blog), February 7.

Marchant, William. 1983. *The Desk Set: A Comedy in Three Acts*. New York: Samuel French.

Metz, Rachel. 2014. "OK, Glass, Find a Killer App." *MIT Technology Review* (blog), January 2.

Mohoney, Gillian. 2014. "Researchers See If Google Glass Can Help Parkinson's Patients." *ABCNews*, April 11.

Monroy, Guillermo L., Nathan D. Shemonski, Ryan L. Shelton, et al. 2014. "Implementation and Evaluation of Google Glass for Visualizing Real-Time Image and Patient Data in the Primary Care Office." *Proc. SPIE* 8935, Advanced Biomedical and Clinical Diagnostic Systems XII, 893514 (February 28). doi: 10.1117/12.2040221 .

Muensterer, Oliver J., Martin Lacher, Christoph Zoeller, et al. 2014. "Google Glass in Pediatric Surgery: An Exploratory Study." *International Journal of Surgery* 12(4): 281–89.

Parkin, Simon. 2014. "10 Breakthrough Technologies: Oculus Rift." *MIT Technology Review* 117(3): 50–52.

Redmond, Eric. 2013. *Programming Google Glass: The Mirror API.* Dallas: Pragmatic Bookshelf.

Rubin, Peter. 2014. "Inside Oculus Rift." *Wired* 22(6): 78–95.

Sawyers, Paul. 2013. "Internet Pioneer Vint Cerf Talks Online Privacy, Google Glass and the Future of Libraries." *The Next Web*, July 12.

Scoble, Robert, and Shel Israel. 2014. *Age of Context: Mobile, Sensors, Data and the Future of Privacy*. n.p.: Patrick Brewster Press.

Shatte, Adrian, Jason Holdsworth, and Ickjai Lee. 2014. "Mobile Augmented Reality Based Context-Aware Library Management System." *Expert Systems with Applications* 41(5): 2174–85.

Shteyngart, Gary. 2013. "O.K., Glass: Confessions of a Google Glass Explorer." *New Yorker* 89(23): 32.

Starner, Thad. 2013. "Google Glass Lead: How Wearing Tech on Our Bodies Actually Helps It Get *Out* of Our Way." *Wired* (blog), December 17.

Thompson, Clive. 2014. "The First Wearables." *Smithsonian* 45(3): 35–37.

Ventura, Marne. 2014. *Google Glass and Robotics Innovator Sebastian Thrun*. Minneapolis: Lerner Pub.

Ye, Hanlu, Meethu Malu, Uran Oh, and Leah Findlater. 2014. "Current and Future Mobile and Wearable Device Use by People with Visual Impairments." Paper presented at ACM CHI Conference on Human Factors in Computing Systems, Toronto, ON.

Yus, Roberto, Primal Pappachan, Prajit Kumar Das, et al. 2014. "Demo: FaceBlock: Privacy-Aware Pictures for Google Glass." In *Proceedings of the 12th Annual International Conference on Mobile Systems, Applications, and Services*, 366. New York: ACM.

Chapter Sixteen

YouTube

Advanced Search Strategies and Tools

Julie A. DeCesare

YouTube and other massively popular video-sharing (like Vimeo) websites have become familiar resources for librarians, instructors, technologists, and information professionals. YouTube is an excellent resource, not only in finding information for patrons, but also for reaching out to them. The YouTube format and visual layout is now familiar to novice searchers and is often a "first stop" for informational, educational, entertainment, and social content. In fact, the main challenge to YouTube is its size and scope. Quality of content on YouTube also varies. The searcher needs to be thorough in their evaluation of the results. Information and meta (visual, media, format, etc.) literacies play a key role in our evaluation of the videos and channels found on YouTube and other video-sharing websites.

This chapter will provide some tips and tricks to navigating YouTube and narrowing search terms, and provide other tools that will help librarians, teachers, and educational technologists utilize YouTube as an important resource in teaching, learning, and research.

A FEW STATISTICS

According to the 2013 Pew online video survey (Purcell 2013), video-sharing sites, mainly YouTube and Vimeo, have been a major driving force in increasing the number of adults who post, watch, and download online videos. The percentage of online adults who use video-sharing sites grew from 33 percent in 2006 to 71 percent in 2011 (Moore 2011). YouTube provides these statistics ("Statistics" 2014):

- There are more than one billion unique visitors each month.
- Over six billion hours of video are watched each month.
- 100 hours of video are uploaded every minute.

VIMEO VERSUS YOUTUBE

Given their popularity, it is important to know about both YouTube and Vimeo, though this chapter will provide more information on YouTube. There are major differences between the two sites that online video searchers and users should be aware of.

Even though it is a popular site, Vimeo is primarily a community of filmmakers, so comments are often more constructive and professional (Larson 2013). The limitations on video length are more open than YouTube's (which limits videos to fifteen minutes). Vimeo has a cleaner layout, and there is more focus on the video. On YouTube, the social components of the site (commenting, sharing, etc.) are as important as the video itself. Vimeo also has limited advertisements, which is a huge difference between the two sites. Advertisements on YouTube are in the thousands. They can be skipped after a few seconds, but still require engagement from the user. Another difference is that Vimeo's video uploader can assign a password to a video. This means that instead of setting the video to "private" and prompting a login to the website, a unique password can be shared with an invited viewer regardless of that person's account status on Vimeo. This password feature allows a video to be private but doesn't require the viewer to set up a Vimeo account. The login is just to see the video. It can be frustrating for a patron or user to be sent a video that requires them to set up an account and may turn them off from using the resource. In addition to the main site, Vimeo also offers:

- Vimeo Video School, a collection of video tutorials and best practices (http://vimeo.com/videoschool).
- Vimeo on Demand, which includes a selection of full-length films (http://vimeo.com/ondemand).
- A directory of Vimeo channels (http://vimeo.com/channels).
- Vimeo Enhancer allows uploaders to modify the looks of the video and add music and audio (http://vimeo.com/enhancer).
- Vimeo partnered with Getty Images to provide a platform for B-roll submissions from Vimeo filmmakers (http://vimeo.com/gettyimages).

YOUTUBE: AN OVERVIEW

YouTube is by far the most popular video-sharing site available (Purcell 2013). YouTube allows users to upload, view, and share video clips. The number of likes and dislikes a video has and the number of times a video has been watched are both published. Unregistered users can watch most videos on the site; registered users have the ability to upload an unlimited number of videos. YouTube's Getting Started Guide provides a step-by-step overview of creating an account, setting privacy/sharing preferences, watching videos, creating playlists, saving videos for later, creating YouTube One Channels, and uploading videos from a variety of devices ("Getting Started" 2014). Uploads are limited to fifteen minutes, but there are a series of steps that can be taken to verify your account in order to allow your channel to have videos longer than fifteen minutes ("Verify Your Account by Phone" 2014).

Verification of account is also based on YouTube's Community Guidelines, which are rules and regulations for uploading and distributing video through their site ("YouTube Community Guidelines" 2014). These are their guidelines for what content can be flagged or removed. For example, sex and nudity, hate speech, dangerous illegal acts, children, copyright, privacy, harassment, and threats are just a selection of topics covered in their guidelines.

YouTube One Channels are a great way to provide visibility and a dedicated space on YouTube for your institution. YouTube Channels are public profiles and playlists of videos from the same provider/individual. All users with an account have the option for a channel. Users can subscribe to channels to be alerted of new content and videos. A couple of examples of very popular libraries with YouTube Channels are:

- New York Public Library
- Library of Congress
- The British Library

UPLOADING AND SHARING VIDEOS: IN BRIEF

Uploading videos to YouTube has become much easier and faster with mobile devices. YouTube offers apps for popular operating systems, such as Android and iOS devices. These apps (YouTube App for Android and YouTube Capture for iOS devices) allow networked devices to "point and shoot," then quickly upload to their YouTube/Google account. Basic metadata, tags, and privacy and sharing settings can all be modified at the point of upload to the YouTube site. YouTube allows for easy linking or embedding into other popular content-management systems, such as Springshare's LibGuides and Wordpress, and learning management systems—Sakai, Moodle, Canvas, and

Blackboard. YouTube provides well-made tutorials and visual aids for their resources.

Video creators can modify their account, upload, and privacy settings by using a desktop browser. These modifications serve as a user's defaults and make uploading from mobile devices even more efficient. Each YouTube user account has options for using e-mail or Multimedia Messaging (MMS) for uploading video. Under YouTube Settings, go to Account Information, then to Mobile Uploads. The account owner will see a unique e-mail address. By e-mailing video files to this unique address, a YouTube account owner can upload videos via e-mail. Video settings and basic editing can be done on a desktop computer if your default settings are not appropriate for the uploaded video. The e-mail option is a huge advantage to get content up and available to patrons quickly.

YouTube has various functions that provide users the ability to comment, annotate, and subscribe to content feeds for individual videos. The browser's video player allows videos to be shared through many social media outlets, such as Facebook, Twitter, and Google+. YouTube also offers closed captioning and options for transcripts on some videos. They are far from 100 percent correct, but this technology has made great strides in the past couple of years. YouTube offers a built-in editing tool as well as options to add captioning and transcript syncing to your personal videos. YouTube support and help pages provide "how-to" info for video creators to provide optimal use of closed captioning, syncing transcripts, and even directional transcripts for users with greater accessibility needs. The YouTube Video Editor allows for slight in-browser editing and modification of an uploaded video, but you need to be logged into your account. YouTube also has a TestTube site (www.youtube.com/testtube) for tools and functionality open to user testing and feedback. This is a great way for video contributors to provide feedback and beta test some of YouTube's new tools.

YouTube has an option to stream live events. This service is offered to users and channels in good standing. "Good standing" means that the channel has abided by community guidelines, has no copyright strikes, and there are no Content ID claims. Content ID is given to copyright holders to easily identify and manage their content on YouTube ("How Content ID Works" 2014). YouTube Live is possible with a laptop and a webcam. The software and hardware needed to optimize the YouTube Live experience, such as an encoding system like Adobe's Flash Media Live Encoder is explained on the site ("YouTube Live Streaming Guide" 2014). A feature separate from YouTube Live is Google+ Hangouts on Air, which enables users to stream their Google+ Hangouts on Air through their YouTube channel. Accounts with a Google+ profile and a YouTube channel in good standing will be able to use this feature ("Hangouts on Air" 2014).

SEARCHING YOUTUBE: SHORTCUTS AND MORE

YouTube is owned by Google, and many Google search shortcuts and strategies can be used to filter and modify results. With YouTube it is very easy to get lost in content. Shortcuts can be used to create more precise and efficient results in the YouTube main search box. Here are a few examples:

- Use quotation marks to search for a specific phrase, e.g., *"pacific octopus"*
- A minus sign (-) can be used to exclude a word or phrase, e.g., *"pacific octopus" -red*
- The *intitle:* command searches for words in a video's title; e.g., *intitle:octopus*
- Adding *HD* to your search query will return high-definition video results, and *3D* will return three-dimensional videos.
- To narrow down to more recent content, add the terms *today*, *this week*, or *this month* to your search string.
- Adding the words *channels* or *playlists* to your search will return results of user channels and playlists.
- If the length of a video is pertinent, add the term *long* to search for videos longer than twenty minutes and the term *short* to return videos shorter than four minutes.
- If you come across a YouTube Channel with a lot of content, you can use a Google *site:* search to narrow results; e.g., *"pacific octopus" site:https://www.youtube.com/user/MontereyBayAquarium*.

By combining some of the shortcuts above, more precise results can be returned. For example, the search string in the last bullet point above retrieves only three YouTube videos from the Monterey Bay Aquarium.

These terms and shortcuts can be seen in YouTube filter options, but adding them at the main search box level is the fastest way to begin narrowing down and controlling results. Searches and settings based on language, country, and content (using Google SafeSearch setting guidelines) are available on the bottom of the YouTube screen. Keep in mind that once an initial search is performed in the main YouTube search box, there is an option to apply filters to results in the top left-hand corner on the returned results page. Filter options include:

- *Upload Date* (last hour, today, this week, etc.)
- *Result Type* (video, channel, playlist, movie, or show)
- *Duration* (short: under four minutes, long: over twenty minutes)
- *Features* (HD, CC, Creative Commons, 3D, Live, Purchased)
- *Sort* (relevancy, popularity, upload date, number of views)

ADDITIONAL TOOLS AND RESOURCES

There are several tools available that can bring YouTube content into context. YouTube Time Machine (http://yttm.tv/) is a really fun tool that allows users to create a historical playlist of videos with a certain date. For example, a search for *1977* returns historical commercial, news, political, and entertainment footage from that year. Overall, it provides historical context and curatorial results to these individual videos that would be lost in the results of a general keyword search on YouTube. YouTube Trends (http://youtube-trends.blogspot.com/) provides a resource to track video and viewership trends. The Massachusetts Institute of Technology Center for Civic Media developed a mapping tool called *What We Watch: Explore Trending YouTube Videos* (http://whatwewatch.mediameter.org/), where YouTube trends can be explored based on international location and date ranges.

Screencast-O-Matic (www.screencast-o-matic.com/) is a very simple and straightforward browser-based screen capture program. Screencast-O-Matic requires no install and provides a one-click interface to begin recording video screen capture (audio is optional). SCOM has a free version with a maximum recording and hosting time of fifteen minutes. The free version allows for direct upload to YouTube, Vimeo, and Google Drive, but there is a watermark. The annual cost of a Screencast-O-Matic Pro account is reasonable at $15 per year and allows for greater functionality like "draw and zoom," password protection at the video level, plus enhanced video and editing tools.

Built to create mashups, lesson plans, and quizzes around TED Talks, *TEDEd: Lessons Worth Sharing* (http://ed.ted.com/) allows users to create quizzes around any YouTube video. Quizzes of up to fifteen questions can be developed for a variety of different answer forms, including multiple-choice and open answers. TEDEd lessons can be made public and shared with the wider TEDEd community. Lessons can also be kept private for use in a learning management system. All users must set up an account, but instructors can share a quiz with students, and all activities, answers, and feedback are recorded for assessment.

Another free site recommended for popular and feature film content is *Movieclips* (http://movieclips.com/). Movieclips is a streaming video library and search engine of clips from popular movies. The great aspect of this site, as opposed to a user-upload site like YouTube, is that the clips and materials have been licensed by Movieclips from larger movie distributors, such as MGM, Sony Pictures, Warner Bros., and several others. All clips are limited to two minutes and thirteen seconds. Clips can be trimmed, resized, shared, and embedded into other sites and resources (like a learning management system). Since the content is licensed and legal, there is less of a chance of the clip disappearing from the site due to copyright infringement, and the website abides by DMCA (Digital Millennium Copyright Act) regulations ("Digital Millennium Copyright Act Notification Policy" 2014).

Clips can be browsed by a variety of categories, for example, actors, directors, theme, mood, genre, prop, and action. Famous quotes, trivia, and additional information are available. Each clip is categorized by Movieclips content curators and is tagged with appropriate metadata. In addition to the website, Movieclips is available as an iPad application, but the clips are in Flash format. Mobile users need to log in with their Facebook account to use the app. Movieclips also provides a great how-to on incorporating its clips into other sites, PowerPoint files, and presentation materials. Movieclips also provides a variety of options for trimming, embedding, and sharing clips. One tip: Movieclips has a YouTube One channel, so you can pull Movieclips into a TEDEd lesson plan via YouTube. It's a great way to develop a media literacy exercise or to get students and patrons more engaged in what they are viewing.

LIBRARIES, YOUTUBE, AND OUTREACH

These are just a few suggestions for utilizing YouTube for outreach, awareness, and community resources.

- Using a screen capture program, like Screencast-O-Matic, record short how-to videos of library databases, catalog searches, or software use. Create a YouTube One Channel, upload the videos to YouTube, and share with patrons. Encourage your patrons to become subscribers.
- If you are reluctant to share the videos publicly, adjust privacy settings to "Anyone with the Link" can view. This would allow you to create a bank of tutorials to share upon point of need.
- Permalinks to YouTube can be quick and easy ways to explain search methods via chat, SMS, or e-mail reference. Plus, the patron has a reference point to go back to for repeat viewing. Use these tutorials to create a "bank" of Frequently Asked Questions for patrons to access themselves.
- If you have access to an iPad, the app *Everything Explained* can add greater functionality. It merges drawing, text, audio, and video capabilities to create a unique instructional video.
- At your next library workshop, take photos and create a promotional video using iPhoto and iMovie to upload to YouTube. A few photos taken throughout the year can create a video "annual report" to share with patrons, board members, benefactors, friends of the library, politicians, and colleagues.
- Create a behind-the-scenes video of prized and unique collections.

- For internal use, private YouTube videos can be a way to document projects and workflows, as well as provide training to new staff and project managers.

YouTube can provide additional visibility and marketing for your library, series, and collections. Libraries will only benefit, when the work we do is seen by patrons and supporters. YouTube provides a free and accessible way to extend our collections and services. Librarians can greatly assist patrons in the search, evaluation, and use of YouTube.

REFERENCES

"Digital Millennium Copyright Act Notification Policy." 2014. *Movieclips.* Accessed March 11, 2014. http://movieclips.com/about/dmca/.

"Getting Started." 2014. *YouTube.* Accessed March 11, 2014. www.youtube.com/yt/about/getting-started.html.

"Hangouts on Air." 2014. *Google.* https://support.google.com/plus/answer/2553119?hl=en&ref_topic=2553242.

"How Content ID Works." 2014. *YouTube.* Accessed March 19, 2014.https://support.google.com/youtube/answer/2797370?hl=en.

Larson, Eric. 2013. "5 Reasons to Choose Vimeo Instead of YouTube." *Mashable.* http://mashable.com/2013/05/30/vimeo-over-youtube/.

Moore, Kathleen. 2011. "71% of Online Adults Now Use Video-Sharing Sites." *Pew Research Internet Project*, July 25. www.pewinternet.org/files/old-media//Files/Reports/2011/Video%20sharing%202011.pdf.

Purcell, Kristen. 2013. "Online Video 2013." *Pew Research Internet Project*, October 10. www.pewinternet.org/2013/10/10/online-video-2013/.

"Statistics." 2014. *YouTube.* Accessed March 11, 2014. www.youtube.com/yt/press/statistics.html.

"Verify Your Account by Phone." 2014. *Google.* Accessed March 11, 2014. https://support.google.com/youtube/answer/4523194?rd=2.

"Video Editor." 2014. *YouTube.* Accessed March 11, 2014. www.youtube.com/editor.

"YouTube Community Guidelines." 2014. *YouTube.* Accessed March 11, 2014. www.youtube.com/t/community_guidelines.

"YouTube Live Streaming Guide." 2014. *YouTube.* Accessed March 19, 2014. https://support.google.com/youtube/answer/2474026?hl=en&ref_topic=2853713&rd=1.

Part III

Networking

Chapter Seventeen

Google Tools and Problem-Based Instruction

Collaborate, Engage, Assess

Janna Mattson and Mary Oberlies

USING GOOGLE TOOLS TO OVERCOME BARRIERS

The very existence of libraries is a response to the need for free public access to information. Barriers to the successful fulfillment of this need include the digital divide, shrinking budgets, and the lack of awareness of library services and resources. In academic libraries, additional concerns include fostering and assessing student learning in a one-shot library workshop, and peer-to-peer collaboration and support while juggling multiple responsibilities. In this chapter we will examine the achievement of three goals using the Google tools Docs (documents), Forms, and Sheets (spreadsheets) to (1) overcome geographic distance, (2) enhance student engagement and foster student-centered learning through problem-based instruction (PBI), and (3) capture assessment data seamlessly.

Google tools can be used to enhance communication and collaboration for users, thus overcoming geographical distance. Web tools provide a better alternative to phone or e-mail contact, as cloud-based documents provide a universal platform, decreasing worries about whether it is necessary to create a Word document versus Linux or corruption of files when downloading and transferring. It is also possible for multiple people to manipulate the document simultaneously, thus providing instantaneous feedback and eliminating much of the time going back and forth with different drafts.

Student engagement and learning increases when it is fostered in an interactive environment. Undergraduates today are digital natives, and integrating technology into library instruction sessions is an effective way to establish an engaging collaborative environment through active learning (Denton 2012; Koury and Jardine 2013). Librarian instructors now have a multitude of Web 2.0 tools to select from and include in their sessions.

Another benefit to using Google tools is that many options are available in a single platform and most users have Google accounts. Even if users do not have a Google account it is still possible to access and use these applications.

David Denton of Seattle Pacific University cites Google Docs as having "the potential to enhance instructional methods predicated on constructivism and cooperative learning" (Denton 2012). Cooperative learning is a crucial element of PBI, and these collaborative efforts and their sustainability between librarians, between students, and between librarians and students depend on tools that are convenient and easy for all to use.

Assessments of library instruction sessions at George Mason University Libraries are being reenvisioned and these PBI efforts are an opportunity to contribute to assessment efforts. Rather than limiting to pre- and post-instruction assessments, using Google tools to create documents and forms to capture student activities provides a more holistic view of what is happening during the session. An added benefit is the option to graphically display student responses from Google Forms, which allows the librarian to check responses quickly and easily. Librarians must begin to think in terms of multiple assessment methods as the need to assess in traditional higher education is becoming increasingly more important, as competition from alternative online learning environments continue to rapidly emerge. In his book *Teaching Naked*, José Bowen advises faculty to "embrace a culture of assessment," as "we will not know if anything we are doing is working" (Bowen 2012, xii). Using Google tools to bolster those efforts will help ease any assessment anxiety.

PBI AND GOOGLE TOOLS IN CONFLICT RESOLUTION

George Mason University's School of Conflict Analysis and Resolution (S-CAR) provides undergraduate and graduate education focusing on the theory, development, and practical facets of conflict analysis and resolution. Student research often examines current events and case studies, applying theory and analyzing causation. In their final capstone course project undergraduates must demonstrate all the skills they have learned. The assignment requires the students to design a research project where they identify a conflict of their choice and apply theory to explain it, evaluate peace-building initiatives (or responses to the conflict), and provide recommendations. Library instruction for this course is important for student success, but it can be difficult to identify where each student is in terms of research ability in a single library instruction session.

After consulting with faculty leading the course, I decided the best way to arrange a library instruction was through group work using PBI. It was important for students to be able to find information, but, more important, to be able to evaluate its relevance and scholarship. PBI places an "emphasis on higher-order critical thinking skills" through the use of collaborative and active learning assignments in which students use their prior knowledge to solve a problem while acquiring navigation and evaluation skills (Cook and Walsh 2012; Yew and Schmidt 2012). Assessment yields the best results when occurring before, during, and after a session, and I found Google Forms and Sheets useful instruments for this. Students access the Google form during class and the Google spreadsheet allows me to create an observation grid. Google forms are an effective way of creating evaluation/assessment tools. They are easy for students to access and fill out, and for me to check. They also provide several helpful ways for me to graphically look at the data and measure how it changes over time. The best part of using Google Forms and Sheets is that it works in real time, allowing me to check responses during the session and assess student progress and determine which groups are struggling, are ahead, or are not paying attention.

Google Forms

Google Forms provides the librarian instructor a useful way to assign tasks that students can complete quickly and easily. Before the sessions, I prompt the faculty to remind students to bring computers and that we will be doing group work during our session. I provide a short URL for the Google form to the students after they have broken into groups. One student serves as secretary who will enter the group responses into the form.

Since groups share their results at the end of the instruction session, I created two forms, each with a different research question. This improved the odds of students learning added skills through other groups' experiences. Since my instruction sessions alternate between lecture and guided learning, Google Forms allows me to test the students throughout the session to see if they are engaging with the lecture. As Google Forms provide an option to include checkboxes, short answer, long answer, or multiple choice, I utilize each option. Questions following a lecture include checkboxes and multiple choice, testing the student on what was covered and asking them to apply it to their research topic. This leads to task completion and providing textual responses that capture citation or student self-assessment data. Section headers are a useful way of arranging the document as they allow me to introduce the topic, the next idea being covered, or provide a prompt for the session activity.

Google forms are a useful way to:

- Intersperse assessment seamlessly throughout the session within a single form.
- Maximize instructional time.

Google forms are multifunctional and can serve as online surveys and worksheets by:

- Utilizing the checkboxes, multiple choice, Likert scales, and text boxes to include quick assessment questions and feedback.
- Utilizing paragraph boxes for longer responses such as providing citations.
- Organizing your form using section headers and page breaks to create "stop" points to correspond with the lesson plan.

Google Sheets

Instruction sessions centered on group activities can easily become difficult to manage, especially when there is a solo instructor. Google Sheets provides a solution. Making spreadsheet grids and tables is a simple process in Sheets because of its format and functions. It is possible to freeze rows and columns to the size needed for content, color them for organizational purposes, and include formulas for adding, subtracting, and averaging values. We used Sheets to create a simple observation grid that served as a checklist. Having the spreadsheet open on a tablet or phone allows the librarian instructor to quickly enter information. If an assistant is also entering observation data, spreadsheets can help the instructor keep tabs on the activity going on in the groups, providing a better understanding of when it is necessary to pull the class together again or redirect groups.

Google Sheets are a useful way to:

- Improve classroom management through checklists and observation grids.
- Organize group activities and student responses for evaluation.
- Visualize data through graphs and summarized responses.

Assessment: Effectiveness and Student Engagement

A benefit to using Google Forms for instruction sessions is that students can move seamlessly through the tasks they are completing. They were using computers to conduct their research, and by having an online form they could simply copy and paste or quickly type their responses. When organizing my lesson plans I found I did not have to budget as much time for each portion, because I no longer needed to wait for students to transcribe answers. This allowed students more time to present their findings and ask questions upon the completion of the assignments. Overall, using Google tools improved instruction time management. Library instruction is often challenging, because generally librarians only get one opportunity to introduce students to library resources and how to conduct research. This makes time management and having an engaging session important for student success and sustainability of skills.

Breaking the form into sections can help control how quickly students move through each aspect of the assignment. It is possible to dictate the next page the form goes to based on certain responses when using multiple choice or list functions. If questions are arranged correctly it is possible to control what the students see. Alternatively it is possible to prompt students to stop at the end of the page, and then have them select the "Next" button to move on when ready. The real benefit of using a Google form as an assignment sheet is that it is so customizable. If it turns out the order of the questions hinders the progress of the session, it is easy to move the questions around, or provide different question types (list, checkbox, etc.).

One potential problem with using Google Forms for these sessions is that students sometimes do not remember to submit the form at the end of the session. If the librarian is not watching the responses coming in at the conclusion of the session this can slip under the radar. To assist with this problem, we found that having two librarians—or at least one other person in the room to do the observations—was helpful. The librarian can then circulate and be on hand for questions and the other person can monitor responses on the Google forms.

Whether it is possible to have another person assisting with observing the groups and monitoring the student responses the decision to use Google Sheets to create an observation grid proved very helpful. If available, an assistant can fill out the sheet and assist with monitoring how the groups are doing. But if working solo, the spreadsheet is useful as it serves as a reminder of which groups need to be checked on. Following the session, the spreadsheet proves a valuable evaluation tool that can be used to improve future lesson plans.

PBI AND GOOGLE TOOLS IN CRIMINOLOGY, LAW, AND SOCIETY

The George Mason University Department of Criminology, Law, and Society's (CLS) mission is to "provide a rich educational experience for our students and to generate research and scholarship" with the goal "to encourage students to go beyond narrow technical topics to grasp the larger role of the criminal justice system in society" (George Mason University Department of Criminology, Law and Society 2014). Since a basic principle of PBI is using real-world scenarios students will likely encounter on the job, it is a teaching method that supports this mission, encouraging students to think beyond the walls of the university. Successful implementation of PBI includes the creation of an engaging problem scenario.

While we have touched on engaging Millennials through the use of technology, the use of a problem to engage the student is founded in John Dewey's philosophy as outlined by these University of South Carolina librarians: "The work of Dewey (1933) provides an early foundation for the concept of inquiry in academic discourse through the development of a six-stage inquiry model. For the inquiry to commence, the individual must be mired in what Dewey calls a 'problematic situation'" (Weaver and Tuten 2014). Using PBI with various Google tools in a library workshop for the undergraduate CLS class "Effective Responses to Crime: Policies and Strategies" allowed for streamlined creation of the problem scenarios; planning; student, librarian, and professor collaboration; student engagement and student-centered learning; and finally various assessments of the workshop.

Google Docs

The development of a PBI workshop began with careful planning and collaboration with the professor. After the students' research topics were forwarded to me, I grouped them by similarity on a Google Doc, created problem scenarios for the student groups to research, and shared the document with the professor for fine tuning. Using a Google Doc gave us a shared point of access without having to e-mail back and forth. With the channels of communication vastly improved, we had more time to focus on the learning experience for the students. I also utilized a Doc with the students to inform them of their team assignments, general research topic, and designated group spokesperson. Sharing the Google Doc also allowed the students to reassign another spokesperson in cases of absence.

Using Google Docs in the Criminology, Law and Society PBI workshop allows for:

- Ease of collaboration with faculty while planning problem scenarios.
- Ease of pre-workshop communication with students.

Google Forms and Sheets

The group work is a crucial element of PBI, as it allows students opportunities for scaffolding as they use their prior knowledge, teaching each other what they know about research in the criminology field. Because the scenarios I created were based on their research topics, each student was a stakeholder in their own learning, which "reflects real-world situations which require team work, collaboration, interpersonal skills and the motivation of various team members towards a common goal" (Cook and Walsh 2012). The morning of the PBI workshop, I e-mailed each spokesperson a Google form that included the scenario and three blank spaces in which to put their citations. The students arrived at the workshop with their groups already formed, alert and ready to work.

After I gave the students a brief introduction to criminology resources available to them from the University Libraries and how to evaluate resources using the CRAAP method (Currency, Relevance, Accuracy, Authority, Purpose), they were ready to work in their groups to solve the problem. Once they chose their citations, they entered them into the Google form that formatted their responses into a Google sheet upon submission. This allowed me to easily assess the quality of their citations using the CRAAP rubric that was also used by my S-CAR librarian colleague. The theme of these transactions, between the faculty member and me, the students and my librarian colleague, is collaboration and connectedness, which was facilitated with Google tools.

Connectedness can be a challenge during a one-shot library instruction workshop. Nevertheless, as Whiteside and colleagues write, a "wide array of media is available to help us enrich our course content; however, it is also important to identify various media and resources aimed at connectedness. Examples include collaboration tools (such as Google Docs), shared calendars, blogs, wikis, online discussions, annotation software, and polling tools" (Whiteside, Dikkers, and Lewis 2014). Using Google Docs, Forms, and Sheets required more work on the front end than a typical didactic library bibliographic instruction session, but time spent on the front end allowed for increased student engagement, active learning, and authentic assessment of student learning by citation evaluation.

Using Google Forms and Sheets in the CLS PBI workshop allowed for:

- A paperless environment
- Ease of organization and access of student responses
- Focus on student learning and collaboration rather than format
- Capturing of student responses for assessment data

LIBRARIAN COLLABORATION AND GOOGLE TOOLS ACROSS CAMPUSES

George Mason University's distributed library and campus system provides ease of access for students and faculty. Subject-specialized librarians are located on each campus with collection development and liaison responsibilities. With a distance of thirty-two miles separating our Arlington and Prince William County locations, collaboration often must occur from a distance. We created multiple types of documents as we collaborated to improve the format of our PBI sessions. We shared lesson plans, handouts, evaluation forms, and assessment tools. In making these documents we utilized Google Drive, Docs, Forms, and Sheets.

We stored all documents associated with PBI activities in a shared Google Drive folder. We also used Google Docs to establish lesson plans, which allowed us to share ideas and make adjustments to improve their effectiveness. We utilized Google Forms in two ways, for assignment completion and assessment. As previously described, the first was as a worksheet for the students to fill out during the session. The benefits to this were allowing us to observe student engagement and as a time-saving tool. Students are definitely comfortable using Google tools, making it easier to focus on working with the students on research techniques rather than teaching them how to use a less familiar system or deal with multiple handouts.

Google Forms was also used for assessment at the completion of the sessions. Having access to all data empowered us to make informed decisions and plan for the improvement of future PBI workshops. We created a student behavior evaluation grid using Google Sheets for use during the PBI workshops. This was another way we assessed student engagement, the data for which was easily accessed and manipulated into graph format. Having access to all the documents we created for the different sessions, and having access to the evaluations and assessments, allowed us to adapt and improve our sessions.

Using Google tools supports librarian collaboration by:

- Sharing resources on an accessible-anywhere platform
- Providing access to a larger scope of assessment data

OUTCOMES, LESSONS LEARNED, AND OPPORTUNITIES

By using Google tools we were able to instantly assess the quality of student work and share that with our respective faculty members. As shared stakeholders in student learning, faculty and librarian partnerships are better supported with Google tools. In addition, assessment of student learning is becoming an increasing priority at university libraries—using Google tools has helped keep us ahead of the curve and has also revealed where improvements are necessary, as any sound iterative assessment process would. For example, in each of our PBI sessions, citation data showed that at least one student group did not meet expectations. Future PBI sessions will include necessary scaffolding by the librarian in addition to hands-off observation.

Google tools also support our peer-to-peer collaboration. We decided to turn our PBI efforts into a poster presentation, continuing to use Google Drive to share outlines, resources, and ideas for the presentation. This helped supplement our collaborative efforts at a distance when it was a challenge to meet face-to-face. We will also use Google tools to support our future shared efforts, including the development of SMART (Specific, Measurable, Achievable, Results-focused, Time-bound) goals for PBI sessions and a problems repository to be shared by Mason librarians who want to conduct their own PBI sessions. Finally, we are exploring peer-to-peer assessment methods, which could take place in conjunction with student observation. Sharing technology such as the Google tools we outlined in this chapter allows busy academic librarians to create a grab-and-go electronic environment in order to bolster collaboration, student-centered learning, and assessment efforts.

In using Google tools our assessment data

- indicated areas of instruction improvement;
- created opportunities for librarian-to-librarian instruction assessment; and
- supported librarian professional development.

REFERENCES

Bowen, José Antonio. 2012. *Teaching Naked: How Moving Technology Out of Your College Classroom Will Improve Student Learning*. San Francisco: Jossey-Bass.

Cook, Peg, and Mary Walsh. 2012. "Collaboration and Problem-Based Learning: Integrating Information Literacy into a Political Science Course." *Communications in Information Literacy* 6(1): 59–72.

Denton, David W. 2012. "Enhancing Instruction through Constructivism, Cooperative Learning, and Cloud Computing." *TechTrends* 56(4): 34–41.

George Mason University Department of Criminology, Law and Society. 2014. "Mission."http://cls.gmu.edu/about-the-department/about.

Koury, Regina, and Spencer J. Jardine. 2013. "Library Instruction in a Cloud: Perspectives from the Trenches." *OCLC Systems & Services* 29(3): 161–69. doi:10.1108/OCLC-01-2013-0001.

Taylor, Laura, and Kirsten Doehler. 2014. "Using Online Surveys to Promote and Assess Learning." *Teaching Statistics* 36(2): 34–40. doi: 10.1111/test.12045.

Weaver, Kari D., and Jane H. Tuten. 2014. "The Critical Inquiry Imperative: Information Literacy and Critical Inquiry as Complementary Concepts in Higher Education." *College & Undergraduate Libraries* 21(2): 136–44. doi:10.1080/10691316.2014.906779.

Whiteside, Aimmee, Amy Garrett Dikkers, and Somer Lewis. 2014. "The Power of Social Presence for Learning." *Educause Review Online* (May/June). http://www.educause.edu/ero/article/power-social-presence-learning.

Yew, Elaine H. J., and Henk G. Schmidt. 2012. "What Students Learn in Problem-Based Learning: A Process Analysis." *Instructional Science* 40(2): 371–95. doi:10.1007/s11251-011-9181-6.

Chapter Eighteen

Group Projects Facilitated by Google Collaboration Tools

Michael C. Goates and Gregory M. Nelson

College students are often required to collaborate on group projects. These collaborative efforts are an integral part of most undergraduate (and even graduate) education, fostering life skills such as teamwork, personal responsibility, and negotiation. However, most undergraduate and graduate students, busy with school, jobs, and personal lives, find that coordinating these projects is a difficult and exasperating task. Finding time to meet can be a challenging aspect of group projects, even when students are on the same campus. This problem is exacerbated for students who are taking online courses and are unable to meet with group members face-to-face.

One solution groups often employ is e-mailing portions of a project back and forth. A distinct disadvantage of collaborating via e-mail is the lack of personal interaction between group members. Gaining a consensus on any topic or problem can take several days to resolve if e-mail is the only vehicle for collaboration. When the entire group cannot participate in an open and timely discussion, the experience may foster frustration and misunderstanding. Further, when groups cannot meet in person, the technological hurdles of e-mail correspondence can impede key processes like brainstorming, a very effective way to generate novel ideas. Exchanging work and ideas via e-mails often results in a disjointed experience and complicates the expression of complex thoughts or ideas. Alternatively, some students use online course management software to collaborate on group assignments; however, many of these programs are rather limited in their collaborative functions.

Fortunately for students, they have alternatives to using course management software or e-mail for collaboration: Google. Google has developed a variety of collaborative tools that are available to students, faculty, librarians, and anyone else working on a group project. In this chapter, we will explain how to leverage Google's collaborative tools to work on a group project. Lastly, we will discuss a hypothetical group assignment to highlight the utility of Docs, Sheets, Slides, Drawings, Calendar, and Hangouts.

GOOGLE PRODUCTIVITY TOOLS

Google has a full suite of free online tools that can facilitate group collaboration via the Internet. Google Calendar, Docs, Sheets, Slides, Drawings, and Hangouts can be used to generate and manipulate information that a single individual creates so that a group can then edit in a synchronous or asynchronous fashion. Group members must access each of these tools through Google Drive, the platform for all Google collaborative tools. Google Drive is freely available to anyone who creates a Google account.

One benefit of using Google Docs, Sheets, Slides, and Drawings is the feature allowing users to upload documents or presentations from the equivalent Microsoft programs directly into the corresponding Google tool. Sometimes the Microsoft program is superior to the corresponding Google tool, but the Microsoft programs are all proprietary and non-collaborative. Once the user fully utilizes the collaborative functions in Google, he or she can again download the file into a Microsoft program to give it a final polish.

Each of these Google tools starts with one person creating an initial file, which he or she then shares with team members via an e-mail invitation feature within the program. Collaborative projects require all team members to have full editing privileges so that they can make edits whenever it is convenient, often reviewing the document at a later time to make their changes and to see the changes others in the group have made. This type of iterative revision has the strength of keeping a single version of a project in a central location so that the project can be changed and ultimately improved. Unlike swapping files and ideas in e-mails, with Google tools, individual "copies" of a project don't exist and therefore do not need to be e-mailed to the whole group for comments, with one unfortunate member then collecting, collating, and finally incorporating everyone else's suggestions. Further, asynchronous editing via Google tools ensures accountability for everyone in the group. Google tools retain a revision history for each project, allowing everyone to see how much each person has contributed to the project. Below are some of the more useful features of Google collaborative tools.

Feature: Sharing

Documents or files that someone creates with Google collaborative tools can be shared with an unlimited number of Google users. The creator of the file simply clicks on the Share button when the file is open and enters the e-mail addresses of desired collaborators. When sharing a document with others, the creator can specify the editing rights for each invitee. The first option gives users full editing rights, which includes inserting or deleting content. The second option limits users to making comments, with no ability to change the file. The third option restricts users to viewing a file, with no ability to comment or edit. Only the owner of a file has the ability to delete a file completely. Google initially grants ownership to the creator of a file, but this creator can transfer ownership to another user at any time. One file can have only one owner at a time.

Feature: Editing

One advantage of creating files with Google collaborative tools is the ability for multiple users to edit the file in real time. A user's changes to a file are all saved instantaneously and are visible to all other users. Each user can view the revision history for a file and, if necessary, revert back to a previous version. When viewing a file simultaneously, each user can observe changes as they are happening.

Using Google collaborative tools also allows users the flexibility of viewing and editing files at their own convenience. Users can edit files directly at any time through Google Drive, or they can edit files collaboratively during a Google Hangouts session. Users with editing rights can view the revision history and attach comments to a file. This allows all collaborators to work effectively, even in an asynchronous setting.

We will briefly discuss the merits of Google Calendar, Docs, Sheets, Slides, and Drawings, followed by discussing the advantages of Google Hangouts as a real-time collaborative space. Below is a detailed description, including potential limitations, for each of the collaborative tools.

Tool: Google Calendar

Google Calendar is an online calendaring and scheduling program that allows users to coordinate group activities. The user who initially creates this calendar can share it with other users by selecting the "Share This Calendar" link under the calendar's drop-down menu.

Benefits of Google Calendar. Google Calendar can be shared with an unlimited number of users. The calendar's administrative level can also be specified for each group member. The creator can grant group members varying permission levels, ranging from full editing and notification capabilities to only being able to see meeting times. When adding items to the calendar, group members can create automatic e-mail and pop-up reminders to be sent to all group members. Calendar is instantly updated, and people can comment on items added to the calendar.

Challenges of Google Calendar. Google Calendar is not integrated into the Google Drive platform, so any calendar must be created independently from other collaborative tools. Also, Calendar is poorly integrated with Hangouts. Users are unable to announce a Hangouts event on Calendar from within the Hangouts interface.

Tool: Google Docs

Google Docs is an online word processing application similar to Microsoft Word that allows multiple people to write, edit, and format text documents.

Benefits of Google Docs. In addition to the features common to all Google collaborative tools, Google Docs also has the added benefit of a relatively simple interface. Most users will find the features of Docs easy to understand and implement, especially if they are familiar with standard word processing software.

Challenges of Google Docs. While Google Docs has many of the same functions as other word processing software packages, it has some limitations. It is difficult to format the document, such as formatting bulleted lists, and setting indents and line spacing. Docs also has no prefabricated document templates. Currently, Docs does not have the ability to easily insert or format citations within the program using a bibliographic management system such as RefWorks, EndNote, or Zotero. Unlike Microsoft Word, Docs does not have a Track Changes feature. The only way to see edits made to a document is through the revision history. The revision history does not allow users to accept or reject individual edits made to a document. Instead, *all* changes made during an editing session are saved as a single extensive change and must be accepted or rejected in their entirety.

Tool: Google Sheets

Google Sheets is an online spreadsheet application similar to Microsoft Excel that allows users to create charts and graphs from existing data.

Benefits of Google Sheets. Google Sheets can easily organize and store data for future use. Sheets supports multiple cell formula functions to enable data analysis and manipulation. Examples of these functions include mathematical equations, statistical analyses, and logical operators.

Challenges of Google Sheets. Google Sheets does have some limitations in comparison to conventional spreadsheet software. Sheets has fewer formatting options for charts and graphs. Users are not always able to change certain features of the chart, such as axis labels or the legend's exact location. Moreover, modifying a chart layout requires some extra work.

Tool: Google Slides

Google Slides is an online application that allows users to create presentation slide shows similar to Microsoft PowerPoint.

Benefits of Google Slides. With Google Slides, group members can upload images into the presentation, including those created using Google Drawings.

Challenges of Google Slides. Although users are able to create professional-looking presentations with this tool, users have relatively few default template styles from which to choose. However, a user can upload additional themes from saved presentations created in programs such as Microsoft PowerPoint.

Tool: Google Drawings

Google Drawings is an online art application similar to Microsoft Paint that allows users to create collaborative online drawings and other creative images.

Benefits of Google Drawings. Users can create freehand drawings or insert various lines and shapes to create more structured drawings. Users can also insert photos, clipart, and other images into a drawing, including images from the Google online library. Once a user completes a drawing, he or she can easily insert it into a presentation or other collaborative document. Users can also publish drawings through a unique URL, making them available to anyone.

Challenges of Google Drawings. The Google Drawings features are rather basic and best suited for simplistic image creation. The drawing tool is somewhat cumbersome, especially for more detailed drawings. The shapes available for insertion are limited; in addition, the formatting options for those shapes and for inserted text are limited.

Tool: Google Hangouts

Google Hangouts is a collaborative online space that allows up to eleven simultaneous users to join a virtual gathering (i.e., a Hangouts session). It is a viable alternative to holding an in-person meeting when everyone cannot meet face-to-face. The session offers two ways for users to interact in a session: a video chat and/or text chat. The video mode enhances the experience with both visual and audio interaction as group members share their thoughts and concerns.

Benefits of Google Hangouts. Real-time interaction within Hangouts is one of the greatest benefits of this tool. Nothing can compensate for live interaction, even if it is through a video call. The online environment of Hangouts also frees up people's schedules so that the group project can move forward. Individuals can participate in group discussion wherever they are as long as they have a dependable Internet connection.

Challenges of Google Hangouts. Internet reliability can be a significant, yet surmountable challenge. If a poor connection exists during a Hangouts session, individual group members can adjust the Internet bandwidth, shut off video cameras, and communicate using the chat feature. It is always a good idea to have a trial run with group members before trying to accomplish any real work. This allows each member to gain experience within the Hangout environment. After the initial session, everyone knows what to expect and the following sessions flow much easier.

Each of the Google products we have already described (Docs, Sheets, Slides, and Drawings) can be displayed in the Hangouts main screen so that each person can see the same images and documents. Any member invited to the Hangouts session can edit these images and documents after obtaining permission from the document's owner. The session virtually provides each member with simultaneous access to the project as if everyone were in the same room, looking at the same document.

The following paragraphs describe the Hangouts screen architecture seen in figure 18.1. The Controls toolbar at the top of the screen is illustrated as Icons A–F. Icon A allows a user to invite people to the open Hangouts session. Once a user selects this icon, Google displays all available e-mail

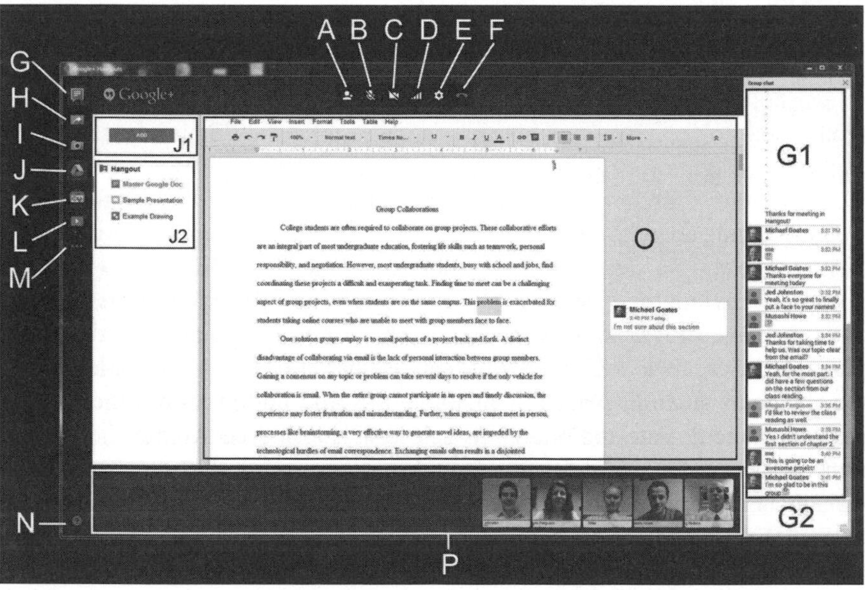

Figure 18.1. Google Hangouts Screen Architecture

addresses or names in a search box and then offers suggestions for potential contacts. After choosing one of the e-mail addresses or names, Hangouts will initiate a message inviting that person to the active Hangouts session.

Icon B allows the user to mute and unmute his or her microphone, and Icon C allows the user to turn his or her camera on or off. Icon D lets users adjust the video quality based on the available bandwidth of a user's Internet connection. The highest bandwidth is the default connection setting, but the user can adjust it lower if needed. Icon E opens the settings function, allowing the user to choose a microphone or camera different from the default setting. Included within the Icon E settings is a sound test to determine if the microphone and speakers are working properly. Icon F lets the user disconnect from the Hangouts session.

The Apps toolbar on the left side of the screen, labeled as Icons G–M, control the basic features of Hangouts.

Icon G allows members to toggle the chat window on the right side of the screen on and off (see Area G1). To use the chat feature, a user types text into the box at the bottom of the chat window (see Area G2). The chat area contains the commenter's name and the comment's entry time, with the comments listed in chronological order.

Icon H lets users activate screensharing. Any Hangouts user can choose to display his or her screen to the entire Hangouts audience. This can be useful for displaying a website or an item on someone's computer. When a user selects Icon H again, the screen will return back to the regular Hangouts session screen.

Icon I, the capture icon, allows members to take a static picture, otherwise known as a screenshot, of what is displayed on a user's screen. Once a screenshot is taken, any of the Hangouts users can view the screenshot. To return back to the Hangouts home location, a user should simply select the capture icon again.

Icon J allows users to link to the files that are on the user's Google Drive. A user cannot display any documents in Hangouts that he or she does not own. For example, only the owner of a Google Docs file can open that document in Hangouts for everyone else to see and edit. Each of the Google tools we have previously discussed can have only one owner at a time, but the material created with these tools can be shared with many people during the Hangouts session. Area J1 containing the Add button allows the user to place documents into the Hangouts selection queue (see Area J2). When a user displays a document for the entire Hangouts group, the actual owner of that document must grant permission for everyone to view the document, even if the user has sharing rights in Drive. This is because Hangouts does not recognize a user with sharing rights, only a user with ownership rights.

Icon K is the Hangouts Toolbox. It contains five tabbed sections that we will mention only briefly. The first section provides the option to display a user's name in the lower third of his or her profile picture in video mode. A user can also add a tagline under his or her name. The second section allows users to turn on a global Hangouts mute, indicating that he or she is currently unavailable. The third section turns on the Comment Tracker application. For more information, check out the website allplus.com. The fourth section is the About tab that connects users to more information about Hangouts. The last section is the Settings tab that allows Hangouts to automatically load previous settings at the start of a video call, to automatically mute all notification sounds, and to automatically load the user's name in the lower third of their picture and tagline from the last Hangouts session.

Icon L lets members connect to YouTube. When a user shows a YouTube video directly within Hangouts, every participant will see the same video and YouTube playlist. When a user skips a video, everyone in the Hangouts session will also skip that video. Only the person initiating the YouTube app will see the search query and results as videos are retrieved. The user can also share the video playlist after the Hangouts session has ended.

Icon M, represented by the triple dots, displays a variety of apps that function in the Hangouts environment; some of the apps are productivity tools and some are just fun and games.

Icon N contains links to Help, keyboard shortcuts, and feedback.

Area O is the main display where documents appear or where the video feed of individuals using video calls appear. This is the working space where members can discuss and edit documents created with the Google tools discussed earlier. One aspect to be aware of is that the document view on each member's screen is static. When someone scrolls through a document, it does not appear to scroll on everyone's screen. If something needs to be edited as a group, then the screenshare feature (Icon H) is the best option. The person who initiates the screenshare is the only one who can edit the document.

Area P is where all Hangouts participants' avatars or live video call feeds will appear. To bring a person's video call into the main screen, a member simply clicks the box of the person in Area P and his or her video feed will populate the main screen. To shift back to viewing the document being reviewed, click the Drive icon (Icon J).

USING GOOGLE TOOLS IN A HYPOTHETICAL GROUP

As an example, let's suppose that a teacher assigns five students to work on a group project. For this project, the instructor requires the group to write a report and prepare a slide presentation to share with the rest of the class. The first step this group might take is to set up a Google Calendar where group members can enter due dates for drafts of the assignment and can schedule meetings, either in person or on Google Hangouts.

After scheduling such calendar events, group members can more effectively collaborate on other portions of their assignment. For the report, a member might create a Google Docs file. While one student drafts the report in Google Docs, other students in the group can be working on graphics for the project using Google Drawings. Other students might also be working on the presentation using Google Slides.

While group participants might initially work on these assignments independently, they could find it more beneficial to meet via Google Hangouts to ensure that all aspects of the assignment are on track. A group member could then send invitations to meet together in Hangouts via the invitation feature in Hangouts itself, an app that links to Gmail, or via the invitation feature in Google Calendar. It might prove very helpful if the students designate one person as the facilitator for the Hangouts session; this facilitator could provide an agenda sent with the invitation and periodically displayed in Hangouts.

During a Hangouts session, the facilitator can call upon each member in turn to lead the discussion on his or her assigned portion of the project using the screenshare feature in Hangouts. Then the group members can discuss the document, spreadsheet, drawing, or slide presentation in real time, and the facilitator can make changes based on everyone's feedback. Differences of opinion can be ironed out in real time because everyone can see the same thing, view each other's faces, and hear each other's voices. This personal interaction, though virtual, is much more likely to allow members to avoid misunderstandings and to come to a consensus on any particular part of the project or the direction of the whole project.

CONCLUSION

Google collaborative tools can effectively improve the flow and final outcome of group projects. This is particularly useful for college students, who often find it difficult to meet in person to complete course assignments. Google Hangouts is a convenient and effective way to utilize each of these collaborative tools, mitigating many common pitfalls of group projects. Users can engage in real-time interactions, combining the benefits of a face-to-face meeting with the convenience of cloud computing. Because these tools are freely available and easy to use, they are useful for a variety of applications, both in academic and nonacademic settings. Although each of these tools has limitations, they grant users far greater collaborative flexibility than more traditional methods.

Chapter Nineteen

Real-Life Experiences and Narratives

Steven Pryor, Therese Zoski Dickman, and Mary Z. Rose

Southern Illinois University Edwardsville (SIUE) emeritus professor Eugene B. Redmond donated his personal collection to SIUE Library and Information Services in 2009. In 2011, a library working group agreed that a series of oral history interviews with Redmond, who was then seventy-two years old, would be a valuable supplement to the Redmond Collection and the associated digital projects. The result, "The Eugene B. Redmond Interviews" project leveraged YouTube as a presentation tool.

BACKGROUND TO THE PROJECT

Redmond has a close relationship with SIUE as an emeritus professor, alumnus, and recipient of an SIUE honorary doctorate degree. While a professor of English at SIUE, Redmond taught African American literature and creative writing. He also began his career as a black studies scholar at the university's "Experiment in Higher Education" in the mid-1960s. The program sought to assist students in nearby East St. Louis in their transition to a conventional college setting. The studies were designed to be relevant to the lives of the predominantly African American students: essentially an early black studies program. Redmond went on to teach black studies courses at California State University before returning to SIUE.

In addition to his career as a black studies professor, Redmond is a poet himself. His poetry career began during the black arts movement, an Afrocentric aesthetic movement of the 1960s that paralleled the political black power movement. Perhaps his most significant contribution to African American poetry remains his 1976 *Drumvoices: The Mission of Afro-*

American Poetry: A Critical History. This definitive literary survey and analysis of African American poetry covers the earliest anonymous spirituals all the way through black arts movement poetry. At the time of this writing, Redmond continues to publish. He also continues to perform his poetry and lecture at poetry conferences.

The Eugene B. Redmond Collection at SIUE, part of the Louisa H. Bowen University Archives and Special Collections at Lovejoy Library, consists of materials accumulated by Redmond over the course of his career. These materials include:

- correspondence and manuscripts
- hundreds of posters, pamphlets, and flyers
- hundreds of books and magazines
- thousands of photographs

To date, the library has created four digital projects exploring selections of these materials. The balance of this chapter describes the process of creating the fourth project, which resulted in the web page "The Eugene B. Redmond Interviews" (SIUE 2014) on the library's website.

CREATING THE INTERVIEWS

Catalog and metadata librarian Mary Rose implemented the Redmond video oral history initiative under the direction of a library working group. Rose had a research interest in the Redmond Collection. She obtained recording equipment to supplement the library's existing equipment using university startup funds for faculty research. She researched oral history methods and consulted with local-area oral historians. She planned the interviews in collaboration with Howard Rambsy II, an associate professor in the English Department and director of the Black Studies Program. Rose and Ginger Stricklin, the library's digital imaging specialist, collaborated to locate a suitable venue for the interviews.

The effort resulted in three two-hour interviews recorded over a five-month period. Rose and Rambsy served as interviewers, and Stricklin shot the video. Rose then transcribed the interviews using Express Scribe Professional. The cost for Express Scribe was approximately $50 including a limited support package. According to standard oral history procedures, the videos were shown to Redmond to review. Minor edits were made and permission was received from Redmond to publish the interviews on the Internet.

LEVERAGING YOUTUBE

Steven Pryor, the library's director of digital initiatives and technologies, investigated technical options for presenting the videos. Three presentation options were identified and evaluated:

- Lovejoy Library utilizes CONTENTdm for several digital collections. CONTENTdm was evaluated as a platform for this collection. Experimentation showed that the CONTENTdm user experience for video files depends greatly on the end-user's software configuration. Factors include which media player is configured to play certain file types and whether videos open in a browser plug-in or an external player.
- The second option was to upload the files to SIUE's web server and make them available for download or embed them in the collection's website. Several copies of each video file would need to be made in various formats to enable reliable streaming across browsers and devices.
- The third option was to leverage the established video distribution infrastructure of YouTube. Lovejoy Library launched a YouTube channel in 2012 to host brief instructional and informational videos and to promote library events and services. The YouTube platform is a robust and reliable host for streaming video on nearly any computer system or mobile device. Due to its extreme popularity, many users of the web are familiar with its interface, options, and player controls. Links to suggested related videos encourage serendipitous discovery of content by users.

The advantages of the third option led Pryor to recommend leveraging YouTube as the video host for the Redmond oral history interviews. The remainder of this chapter describes the steps in this process.

Step 1: Reviewing the YouTube "Terms of Service"

The YouTube "Terms of Service" were reviewed to ensure that the library and university were not relinquishing any unnecessary ownership rights to these important materials. Indeed, the "Terms of Service" web page states in part 6C, "For clarity, you retain all of your ownership rights in your Content." YouTube is granted several rights related to duplicating, distributing, and performing the "Content" in any location and any format "in connection with the Service and YouTube's (and its successors' and affiliates') business." Further, it is specified that these rights terminate within a "commercially reasonable" time after deletion if the owner decides to remove the videos ("Terms of Service" 2010). It was agreed that these protections were adequate and appropriate for the sharing of this content.

Step 2: Configuring the Account to Accept Long Videos

The next step was to prepare the library's YouTube account to accept videos of the length necessary for the oral history interviews. The three videos were each nearly two hours in length, far beyond YouTube's default fifteen-minute time limit. It is relatively easy to "verify" a YouTube account. This process removes the time limit assuming the account is in good standing regarding copyright claims and community guidelines. Verification requires providing a mobile phone number to YouTube, and confirming a code sent to that number via text message. Once the code is confirmed, the user is able to upload longer videos.

Step 3: Uploading the Videos to the Account

The edited videos were then uploaded to the library's YouTube account through the "Export" function of Final Cut Pro. Adobe Premiere, Final Cut Pro, and similar software suites can usually export directly to YouTube by entering the account details, video quality settings, and other basic information in a software menu. Videos can also be saved to a file and uploaded via the YouTube website.

Step 4: Determining the Appropriate Privacy Level

When the video is uploaded, content owners can specify the initial privacy level:

- private: visible to the account that uploaded it and other Google accounts specifically granted access to that particular video
- unlisted: not retrieved by searches or included in suggestions, but may be seen by anyone who has the link
- public: fully indexed and searchable, viewable by anyone, and may appear in search results or related videos suggestions

When uploading videos, the authors suggest setting the initial privacy setting to "private" so that the video's metadata can be configured before it is exposed to the public.

Step 5: Adding Metadata

Metadata for a YouTube video consists primarily of the title, description, category, and tags. Additional data may be added such as recording location and date, annotations (pop-up text or labels over the video content), and closed captioning options. In choosing the title and description for the videos, care was taken to be both descriptive and mindful of the effect the chosen data will have on the visibility of the content. In the description field, the web address of the corresponding digital collection page was placed in the first line. The YouTube interface only shows the first three lines of the description without clicking "Show more," and those first few lines may be all that most users see. The rest of the description contains the topics in the particular interview, the date the interview was conducted, and the names of the interviewer(s) and camera operator for the session. YouTube tags provide the rest of the data that affects search indexing. These were chosen based on broad content descriptors and the suggestion of the YouTube software: for example, "Poet (Profession)" and "Poetry (Literary School)." Tags are not generally visible to end users but will link the video to highly relevant content elsewhere on the site and increase the chance that someone will find and watch it.

Step 6: Adding Captions

Closed captioning is important not only for accessibility reasons; captions are searchable as well. When any video is uploaded to YouTube, speech-to-text processing occurs that generates an automatic closed caption accompaniment to any speech detected in the video. The reliability of this method varies greatly depending on the speaker and the content. YouTube provides additional options that may be used to improve the results.

- The owner of the video may view the automatic transcription and manually correct mistakes. For short videos, this is often an easy way to achieve an accurate transcription. At nearly two hours long each, the oral histories would require a significant amount of time to check and correct the automatic transcription.
- YouTube allows content owners to upload transcript files, either as a plain text file or a time-coded file following specific caption file format conventions.

Since Rose had already transcribed the interviews to accompany the digital collection, the second option was utilized. Because the time-coded files require each line to be coded with a start and end time (also very time-consuming), we started with the plain text file option. With this approach, YouTube again applies its speech recognition algorithms to the video. However, instead of inserting the stream of words it thinks it recognizes, it attempts to match the speech to the words in the transcript text. This ensures that only text from the actual transcript, not nonsensical phrases, appears in the captions. Since captions are searchable, this means that the video only appears in correct and relevant search results pages. A significant problem with this method for this project, however, is that for videos longer than one to one and one-half hours the timing of the captions accumulates significant errors late in the video. Like transcription errors, this can also be corrected manually with an investment of time.

CONCLUSION

The interviews are easily accessible to the viewer on the library website via the embedded YouTube platform. It is hoped that making the videos public on YouTube will also enhance the visibility of the interviews. YouTube provides a sophisticated "Analytics Dashboard" to assess the success of the videos in attracting attention. An overview report provides graphs and numbers for the number of views, likes, comments, shares, and the estimated minutes watched for the entire account or for specific videos. More specific reports provide additional details. From this data, an uploader can determine whether a video is reaching its target audience and whether the metadata and keywords are effective in driving search and relevance-ranking traffic (serendipitous discovery) to the content.

Redmond discusses many of his life experiences in these oral history interviews. He talks about attending SIUE and his mentors. He discusses writing his critical history of African American poetry and his relationships with other prominent African American writers. He performs and analyzes his own poetry and even demonstrates the yonvalou dance. The video interviews preserve Redmond's life experiences and narratives as only he can convey them, enhancing the Redmond Collection and its growing digital presence.

REFERENCES

SIUE (Southern Illinois University Edwardsville). 2014. "The Eugene B. Redmond Interviews." Library and Information Services. Accessed March 28, 2014. www.siue.edu/lovejoylibrary/tas/EBROH.htm .

"Terms of Service." 2010. *YouTube*. www.youtube.com/t/terms .

Chapter Twenty

Using Google Sites to Create ePortfolios for Graduate Students as a Means to Promote Reflective Learning in the Development of Dissertation Topics

Jesse Leraas and Susan Huber

The use of electronic portfolios (ePortfolios) in higher education is by no means a new phenomenon. In fact, educators have been implementing this educational tool for myriad reasons from assessment to accreditation for many years (Papp 2014); 77 percent of the member institutions of the American Association of Universities have implemented ePortfolios in some manner (Mayowski and Golden 2012). Although many educational institutions have found methods to leverage this technology, we propose a new use for ePortfolios: employing them to capture learner artifacts as a prelude to writing a doctoral dissertation. Students who are able to collect course documents early in their academic careers will find a rich set of data from which to inform future dissertation topics. We chose Google Sites for our ePortfolio platform due to its ease of use, flexibility, dynamic qualities, and cost.

THE PROBLEM STATEMENT

The journey's end for any doctoral student is the culmination of the accumulated knowledge gained through graduate coursework and experience, which results in a doctoral thesis or dissertation as part of the requirement for obtaining a terminal degree. Unfortunately, all too often students have difficulty developing a cogent, researchable topic for their dissertations. Students who come to the dissertation phase of their doctoral program but are unable to complete it are titled All But Dissertation or ABD. Unbeknownst to many students, a vast, rich collection of topics may lie within the artifacts they have created throughout their previous classes; however, if students are unaware of this content, they cannot draw upon it. As many students understand, a poor system of organizing data leads to many lost resources that would be useful for future reference. We suggest that students who are able to look back on the work they have conducted will find repeated themes, which will help them develop a "doable" topic that will lead to the successful completion of their dissertation.

THE PROJECT

This project originated as a collaboration between an adjunct professor of education who teaches the doctoral writing and research course and a librarian who co-teaches the course at a small, Midwestern university serving approximately fifteen hundred students. We have developed a program that integrates ePortfolios into the graduate students' writing course to promote reflective learning and to create a repository for students as a means to inform their dissertation topics. For this project, the use of Google Sites for ePortfolio creation seemed the most logical choice because it possesses many beneficial qualities.

First, it is easy to set up an ePortfolio to act as a repository for documents created by the learner. The platform allows students to choose from various pre-created themes, which is perfect for first-time users. Google Sites offers free space, and it is easy to edit. Second, learners who develop and maintain an ePortfolio are essentially practicing reflective learning. Reflective learning is a process that allows the learner to take a metacognitive approach to the learning experience. Google Sites allows learners to access their website at any time from multiple devices, which promotes access for anytime reflective learning. Third, the integration of ePortfolios introduces students to Web 2.0 technologies, which are becoming more of a necessity to remain productive in our society. It is important for our graduates to develop competencies with social media technologies because of the growing presence of those technologies in academic settings (Lenstrup 2013). Google Sites and other Google apps present some great Web 2.0 uses. Furthermore, students are introduced to the concept of cloud computing technologies.

EPORTFOLIOS

Portfolios are a valuable educational tool because they constantly grow and change with the learner. According to Gambino, the integration of ePortfolios facilitates a student's ability to create a learning experience with meaning and can impact campus culture if incorporated across multiple disciplines (2014). One advantage over traditional portfolios is that ePortfolios afford the learner greater variety of materials and the ability for the learner to manipulate them. For example, ePortfolios include such learning artifacts as papers, presentations, and videos. Typically, portfolios are created in an academic environment and can be used to increase the student's active learning skills (Ambrose, Martin, and Page 2014). The writing and research course has implemented Google Sites for students to use as they gather documents that will lead them to uncover strong research topics for their dissertation.

THE DISSERTATION PROCESS AND THE REFLECTIVE LEARNER

In the process of writing a dissertation the first objective is to determine an appropriate topic. Most important, students must find a topic they can be passionate about and can maintain an interest in for an extended period. The topic will be one that they will become expert in and one that will allow them to use their expertise to pursue career objectives; therefore, the topic selection is an extremely important decision. After the topic has been selected, the research step begins. If the topic is selected early in the student's education process, the research can take place during classes so that by the time classes are completed, much of the research toward the dissertation can be completed.

Reflective Learning

Traditionally, reflective learning occurs during the integration of new information. Following the intake of information, a period of reflection is essential to allow for synthesis. Students can then sort through the new knowledge and apply it to their personal experiences, which allows for the actual learning to occur. Experiential learning allows the learner to skip over some of the reflection in learning, as the actual experiences associated with the learning occur within the same process. When these reflective learning opportunities are woven within the course curriculum, students' ability to think critically about the assignment and to develop a professional identity is improved (Rafeldt et al. 2014).

IMPLEMENTATION PLAN

The implementation of ePortfolios into the research and writing course is aligned with the learning objectives within the eight modules of the course. In our first class session, we discuss the history and importance of ePortfolios, explain how we will implement them in the course, and communicate the learning objectives. We have created six lessons based on Modules 1 through 8.

Lesson 1: Setting up a Google Sites Account

The first objective is to create a Google Apps account and to develop the main page of Google Sites. Students are directed to create a personal mission statement and to add a photo. Following are the step-by-step directions for Lesson 1.

Create an Account

1. Create a Google Account (or use an existing Google account).
2. Log into Google Sites.
3. Click Create.
4. Give your site a name with your name in it.
5. Scroll through the pre-created templates and select a theme.
6. Click the Share button and select Anyone with a Link. Click Save.
7. Click the back button found on the left-hand side of the screen.
8. Click the Edit page button. Click Layout.
9. Choose the two-column option with header and footer layout.
10. Click the header and add a short introduction.
11. Click in the left-hand window and answer the four questions:

    a. Why are you pursuing your chosen field?
    b. On what topic are you planning to conduct your research?
    c. What do you hope to achieve with your dissertation?
    d. Why are you passionate about the topic?

Add a Personal Photograph

1. Click Edit Page.
2. Place your cursor in the right-hand column.
3. Click Insert.
4. Click Image.
5. Click Upload Images.
6. Under the Libraries folder, click Pictures. Select a professional photo of yourself.
7. Click OK.
8. Click the S or M on the image toolbar. Adjust left, right, or center depending on personal preference.
9. Click Save.

Lesson 2: Creating a Journal

Lesson 2 asks the student to create a learning journal for the purpose of reflection on the topic and to add two relevant posts.

Add a Journal/Field Notes Page

1. Log into Google Sites.
2. Click on the New Page icon.
3. Name the page "Journal."
4. Change the template to Announcements.
5. Leave the location as top level.
6. Click Create.

Add a Reflective Post

1. Click New Post.
2. Add a note regarding something you learned in class or from the readings.
3. Change "Untitled Post" to something more fitting to your content.
4. Click Save.
5. Repeat the process so that you have two entries.

Lesson 3: Add Research Questions Page

Lesson 3 requires the student to add a page titled Research Questions and upload a Word document.

Add Research Questions

1. Log into Google Sites.
2. Click on the New Page icon.
3. Name the page "Research Question."
4. Change the template to Web Page.
5. Leave the location as top level.
6. Click Create.
7. Copy and paste your research question and relational research questions into the main window.
8. Click Save.

Upload a Document

1. From the Research Question page, click the "Add Files" link.
2. Find your Research Assignment and select.

3. Click Save.

Lesson 4: Adding a Literature Review Page

For Lesson 4, students are required to create a Literature Review page as well as subpages where they will upload their journal citations.

Add Literature Review

1. Log into Google Sites.
2. Click on the New Page icon.
3. Name the page "Literature Review."
4. Change the template to List.
5. Leave the location as top level.
6. Click Create.
7. Find the Create Your Own option and click Use Template.
8. Click Add Item.
9. Enter your first citation.
10. Click Save.
11. Repeat these steps until all citations are uploaded.

Add Subpages

1. Click on the New Page icon.
2. Name the page one of the keywords or terms used informing your research question.
3. Change the template to List.
4. Select the Put Page under Literature Review option.
5. Click Create.
6. Find the Create Your Own Template and click Use Template.
7. Click Add Item.
8. Enter the citations that pertain to that individual key word or term.
9. Click Save.
10. Repeat until all four citations are uploaded onto a specific page.

Lesson 5: Adding a Survey

Lesson 5 requires students to create a page with a demographic survey. This assignment allows the students to become familiar with survey creation. Google provides a free alternative to the more popular, proprietary survey creation websites such as Survey Monkey.

Create Survey

1. Log into Google Drive.
2. Click Create and choose Form.
3. Title the document "Demographic Survey."
4. Choose a theme from the templates listed and click OK.
5. Create a survey using the three questions from the *Demographic Survey Information for Lesson 5* document. All questions are multiple choice and should include "other." After adding an item, click Add Item. When you are finished, click Done.
6. Click View Live Form.
7. Edit if necessary.

Add Survey to Google Sites

1. Log into Google Sites.
2. Click the New Page icon.
3. Title the page Survey.
4. Change the template to File Cabinet.
5. Leave the location as top level.
6. Click Create.
7. Click the Edit Page icon.
8. Click Insert from the toolbar.
9. Click the Google Drive icon. Click Form.
10. Click on your Demographic Survey. Click Select.
11. Click Save.

Demographic Survey Information for Lesson 5

What is your age?

18–24 years old
25–34 years old
35–44 years old
45–55 years old

What is your ethnicity?

African American
Asian or Pacific Islander
Hispanic or Latino
Native American or American Indian
Caucasian American
Other

What is the highest degree or level of school you have completed?

High School graduate or GED
Associate degree
Bachelor's degree
Master's degree
Doctorate degree

Lesson 6: Adding a Presentation Page

Students are required to add a presentation page and upload a PowerPoint presentation.

Preparing PowerPoint for Upload

1. Open your PowerPoint presentation.
2. Click on an image in your PowerPoint. Click Format. Click Compress Pictures. Select E-mail, deselect Apply Only to This Picture, and click OK.
3. Save it to your desktop.
4. Log into Google Drive and click Settings.
5. Click the drop-down menu next to the Settings icon and move your cursor over Upload Settings.
6. Select both the Convert Upload Files to Google Docs Format and Confirm Settings before each upload.
7. Click Create. Click Folder and title it W7000. Click Create.
8. Click the folder and click Upload. Click Files and find your PowerPoint document.
9. Click Start Upload.

Upload a PowerPoint Presentation

1. Log into Google Sites.
2. Click the New Page icon.
3. Title the page W7000 Presentation.
4. Leave the template as Webpage.
5. Leave the location as top level.
6. Click Create.
7. Click the Edit Page icon.
8. Click Insert from the toolbar.
9. Click the Google Drive icon. Click Presentations.
10. Click the presentation you want to upload and click Select.
11. Uncheck the Include Border and Include Title Boxes. Change the size to Medium or Large and click Save.
12. Click Center and Save.

FURTHER IMPROVEMENTS

Facilitators in higher education are charged with constructing occasions for learning that allows students to grow; therefore, it is imperative for those facilitators to adopt the practice of reflective learning as a means to improve course content (Loughran 2013). We are constantly looking for opportunities to improve the learner experience. Here are some of the ways we are working to improve the course over time.

Soliciting Feedback

We solicit summative feedback from the learners. This is valuable information for modifying future class sessions to meet learner-specific requirements. Google Forms is an easy tool to create an online survey for students to express their opinions about the course and suggest future improvements. We are invested in ensuring that students are utilizing their ePortfolios, so we send out regular follow-up e-mails at the beginning of the semester to former students of the class requesting feedback and inquiring about possible problems the student might have with the technology.

Maintaining Lines of Communication

The instructor and librarian maintain regular meetings to discuss possible pedagogical and technical changes to the course. At each meeting, we discuss strengths and weaknesses of the current technology as well as introducing any new technology we feel is relevant.

Growing Campus Support Infrastructure

Although the librarian is the main resource for technical questions, students may seek out assistance from other campus entities. Faculty and students may encounter obstacles with this new technology (Malik et al. 2014). For this reason, the facilitators of the course will inform library staff, Academic Resource Center members, and Student Services.

Keeping Up with the Technology

Google is constantly changing and improving its platform. The facilitators set aside time on a regular basis to tinker with the technology. It is important that we remain current with improvements and other competing products.

Moving from Private to Public Settings

Although we discourage students from making their ePortfolio available publicly on the Internet, we do feel that their portfolio presents a good tool for them to publish other learning artifacts and to create a web presence. At that point, we encourage students to adjust their privacy settings and to develop a custom domain. We are actively working on putting together protocols and lists of best practices to share with students regarding this process.

CONCLUSION

Integrating ePortfolios across the curriculum can foster a positive impact on campus culture by improving the student experience and affecting student retention (Gambino 2014). We present a ePortfolios as a means to enhance completion efforts by equipping graduate students with the tools and training necessary to complete their dissertations. In order for ePortfolios to be successful, three criteria, or changes in "responsibility," must be achieved: the faculty must be part of a cross-curriculum approach, student support services as well as administrative departments must become involved, and students must play a greater role in the learning process at a curricular level (Cambridge 2012). Although incorporating ePortfolios across the curriculum can be a monumental task, it is a worthwhile endeavor, as it promotes student growth through critical thinking and reflective learning.

REFERENCES

Ambrose, G. A., Holly E. Martin, and Hugh R. Page Jr. 2014. "Linking Advising and E-Portfolios for Engagement: Design, Evolution, Assessment, and University-Wide Implementation." *Peer Review* 16(1): 14–18. http://search.proquest.com/docview/1532665149?accountid=34899 .

Boulton, Helen, and Alison Hramiak. 2012. "Writing in the Virtual Environment." In *Writing in the Disciplines: Building Supportive Cultures for Student Writing in UK Higher Education*, edited by Christine Hardy and Lisa Clughen. Bradford, UK: Emerald Insight.

Cambridge, Darren. 2012. "E-Portfolios: Go Big or Go Home." *EDUCAUSE Review*, March 21, 47(2): 52–53. https://net.educause.edu/ir/library/pdf/ERM1226.pdf .

Flanigan, Eleanor J. 2012. "Electronic Portfolios in Business Schools." *Business Review, Cambridge* 20(2): 170–75. http://search.proquest.com/docview/1238684617? accountid=34899 .

Gambino, Laura M. 2014. "Putting E-Portfolios at the Center of Our Learning." *Peer Review* 16(1): 31–34. http://search.proquest.com/docview/1532664210?accountid=34899 .

Lenstrup, Christine. 2013. "Social-Media Learning Environments." In *Learning in Higher Education: Contemporary Standpoints*, edited by Claus Nygaard, John Branch, and Clive Holtham. Oxfordshire, UK: Libri Publishing.

Loughran, John. 2013. "Stepping Out in Style: Leading the Way in Teaching and Learning in Higher Education." In *Pedagogies for the Future: Leading Quality Learning and Teaching in Higher Education*, edited by Robyn Brandenburg and Jacqueline Z. Wilson. Rotterdam, Netherlands: Sense Publishers.

Malik, Savita, Alycia Shada, Ruth Cox, Maggie Beers, and Mary Beth Love. 2014. "Portraits of Learning: Comprehensive Assessment through E-Portfolios in the Metro Academics

Project." *Peer Review* 16 (1): 27–30. https://login.libproxy.edmc.edu/login?url=http://search.proquest.com.libproxy.edmc.edu/docview/1532663447?accountid=34899

Mayowski, Colleen, and Cynthia Golden. 2012. "Identifying E-Portfolio Practices at AAU Universities." *EDUCAUSE Center for Analysis and Research*, June 14. Accessed April 21, 2014. https://net.educause.edu/ir/library/pdf/erb1206.pdf.

Papp, Raymond. 2014. "Assessment and Assurance of Learning Using E-Portfolios." *Journal of Case Studies in Accreditation and Assessment* 3: 1–6. http://search.proquest.com/docview/1518677625?accountid=34899 .

Rafeldt, Lillian A., Heather Jane Bader, Nancy Lesnick Czarzasty, Ellen Freeman, Edith Ouellet, and Judith M. Snayd. 2014. "Reflection Builds Twenty-First-Century Professionals." *Peer Review* 16(1): 19–23. http://search.proquest.com/docview/1532664464?accountid=34899 .

Chapter Twenty-One

Using Google+ for Networking and Research

Felicia M. Vertrees

Google+ has a robust community of users who are regularly working together and sharing information. Matt Cooke, Google+'s product marketing manager, says the platform now boasts some five hundred forty million active users (Benady 2014). Many users are marketing businesses, providing services or products, researching, teaching, and networking. About 40 percent of marketers use Google+, 70 percent are interested in becoming more adept at using it, and 67 percent say they will increase their usage in the future (Jain 2014). Many educators have also started using Google+ to help improve student collaboration skills, do research on projects, and strengthen the bond between student and instructor (Erkoller and Oberer 2013). According to Jain (2014), many libraries now use Google+ to market their services or events and provide information to patrons.

While Google+ doesn't have the activity level of Facebook, it does have a vibrant community of dedicated users. Why should one join Google+? For starters, Google+ gives you the ability to create groups of people who share your interests. For example, if you're an educator interested in researching information literacy, you could use this platform to start a group of like-minded educators to discuss your research findings. If you own a small business and want to move your marketing efforts online, you could create a page that reflects your brand and then form connections with other businesses in your industry (Jain 2014). In essence, you want to network and research with others who share the same interests. That way, you can share information that might help your business or educational institution. Google+ is unique because it allows you to create groups geared toward a particular task and invite other users to join that group. Because it is a global platform, you can realistically network with someone halfway around the world. For example, you want to schedule a meeting with a fellow researcher, but that person lives across the country. To remedy this, you could schedule a virtual meeting using Google Hangouts. Google+ also allows you to share information with select people within a group you create instead of having to share every post with every connection (Balubaid 2013). This way, you can communicate with each individual group, using messages that are specific to that group's interests and needs.

Using a social media platform such as Google+ also allows users to build trust and create diverse working relationships. You can explain to a particular audience why you think the information you are communicating to them is important and ask them for their opinion on that information. A dialogue is established, creating more users who will follow your posts and comment on them (DeLoma 2012). Small business owners also can learn from larger corporations who have embraced social media platforms like Google+. For example, Home Depot announces upcoming do-it-yourself events that will be held at its stores. Associates monitor the store's social media feed and field questions or provide useful information to consumers (Schultz and Peltier 2013). In fact, out of the 177 companies on the 2013 *Fortune* 500 list, 35 percent actively use Google+ (Barnes and Wright 2014).

Whether you're a library or a small business using traditional marketing methods, the costs can be detrimental to your bottom line. One of the reasons organizations have turned to social media like Google+ is that it saves time and is cheaper than traditional marketing techniques. You can immediately reach a larger audience when you post a message, and you have the ability to craft that message in a way that is creative and timely. By posting to users on Google+, you build a relationship with those users. When they respond to a post or share it, your message grows as more users see it and share it with others (Jain 2014). According to Agence France-Presse (2013), having a robust Google+ account actually helps you rank higher in Google searches.

Let's talk about the process you should follow to successfully use Google+. Before you start using any social media, have a goal in mind. For example, if you're a school administrator, create a plan for how using this platform will benefit students and staff. Decide on the purpose of your account. Are you marketing products or services, providing information, or networking? (Jain 2014). If you plan on using the platform to help your students become better researchers, refer to the AASL Standards for the 21st-Century Learner. These standards state that students need to become adept at sharing information through social networks and information tools (Houston 2012). Educators should also investigate Google Apps for Education. While Google+ isn't included in this list, it fits in nicely with the other apps and would be easy for students to learn.

Before you create your profile, write down any information you think is essential to helping others understand your or your organization's purpose and interests. Use this information as a guide when you're creating your profile and when looking for others you might want to follow. Think about mocking up a Google+ page to show your coworkers or manager. Having a plan is also helpful when creating a personal page you will use in a professional capacity. Be aware that, if you're considering using Google+, like most social media, it is restricted to ages thirteen and up (Fredrick 2013). Once you have written down all of the information you want to include in your profile, it's time to create your Google+ account. Use the following steps to set up a personal, business, or academic account (Hines 2012):

- Type Create Google+ Account and follow the links to learn how to set up your account.
- If you have a Google account, sign in. If not, create an account. Create a separate account if the account will be shared with others. Add a photo to the page. If you have other social media accounts, use the same photo to visually link it to those accounts. This consistency allows you to create a unique brand.
- Look for people you know who have a Yahoo! or Hotmail account.
- Look for people you know who are on Google+.
- Find others who share your interests. This is where you can find users who are working on similar research, business owners in the same industry as you, or educators who are teaching using Google+.
- Fill in the rest of your profile information so others know why you're trying to connect with them and why they should connect with you.
- Create an introduction. Talk about you or your organization. Explain your interests and let others know what you hope to accomplish.

Once you've created your account, personalize it with photos and information about your organization (Jain 2014). For example, if your business sells customized baby strollers, have photos of your product on your page along with background colors and images that enhance your product, service, or brand. If you're a library or college, your page should include photos that relate to your institution. Those photos could include your mascot, a recognizable statue on campus, or even students and faculty. Educational institutions, like businesses, need to brand themselves. Whatever message you want to convey, make sure that your images, colors, and text support that message. Start small because you can always add more later (Jain 2014). For example, if you're a small business, you might want to highlight a particular product or service and add other products along the way. This will give you time to see what's working and what's not and tweak your page accordingly. See if people are following you and commenting on your posts. While it's tempting to create your account and start posting right away, take the time to craft informative and relevant posts. If you're a library, you might concentrate on events or library resources (Jain 2014). When posting to your account, stay away from formal speech. This way, users are more likely to engage you and comment on your page. In addition, while you don't have to post every day, you do need to post regularly so the information on your page doesn't get stale. Once you get more comfortable with Google+, create channels for specific interests. For example, if you're a university, create channels for the different colleges and the library. Consider featuring some of your users, which will instantly personalize your page (Jain 2014).

As I mentioned previously, you can create circles that include people who share your same interests. Your circles can only be seen by you and allow you to share posts only with those users, or even a smaller group within that circle. This is different from Facebook, which allows you to make lists of friends. For example, maybe you want to network with other users interested in environmental issues or you're a student who wants to collaborate with students from other cities. You can create a circle to exchange information on your research. Only those users would see the posts and be able to respond to them. Circles are also a great way to keep up with users who share similar research interests. By adding them to your circle, you can follow their posts. Circles can be any size that you choose, so you might have several circles with just a few people in them. If your networking or research group includes only a few people, then create a small circle to share information only with those people. The goal is to create circles that fall into different categories that interest you. You could have circles for each research topic and specific networking needs ("50 Great Google+ Tips" 2012). Instead of simply searching Google+ for topics that interest you, you can start creating circles with users you want to follow.

Let's look at a real-life example of virtual collaboration using Google+. Washington County Schools, in North Carolina, are encouraging students to use Google+ to collaborate on projects with students in Texas. By introducing students, in essence, to the world, students learn that their work has a much broader influence than just in their local communities. They actually can make a global impact. Mastering this technology also prepares them for using different technologies in college (Roscarla 2012).

Google+ Communities is a feature that was launched toward the end of 2013 and could give users access to global information and collaboration. These differ from circles. Circles are your own personal way to organize connections and communicate with a select group of users. Communities are similar to a forum, where everyone can post items to discuss. These groups can be open to everyone or by invitation only, and all of the information stays in one place (Chen 2013). For example, you're working on a grant and you're interested in creating a community to talk about research, divide up responsibilities, and keep each other up to date on the process. You could create the community and post drafts, research articles, or videos that might be of interest to community members. Anyone can join the discussion and post information. You can also look for other Google+ users who might be interested in this topic and add them to a community or a circle on your topic.

Aside from the appeal of using Google+ for outside networking and research, the software also works well within an organization. For example, if your library works with other departments or colleges on your campus or in your library system, this is a great way to share information and network internally. By having different entities within your library create Google+ accounts, every account user will be able to share information using applications such as Google Hangouts, Google Drive, Google Calendar, and YouTube.

Let's look at another way to do research. For example, you're working within a circle of users in Google+ and you would like to discuss a research idea. Google Hangouts allows you to immediately start a virtual video meeting or schedule one anytime in the future. It also allows you to record this hangout and post it on YouTube. This is called a "Hangout on Air." Keep in mind that, if you plan a regular Video Hangout, the meeting won't be recorded (Stevens 2013). These Google Hangouts have many bells and whistles to create a robust virtual meeting. You can watch YouTube videos, collaborate on a document in real time using Google Drive, or even share your screen with the other Hangouts participants. If you need to organize your Hangouts, you can also create an event. Creating an event allows participants to share photos and add the event to their calendar. Keep in mind that you would still need to start the Hangout at the scheduled time. When you schedule an event, Google Hangouts simply sends out an invitation when you create the event and a reminder shortly before the event is scheduled to start. Find hangouts to attend or view past ones by using the search box at the top of Google+ and entering your search terms.

Starting a Google Hangout is simple. Remember you can either start your Hangout immediately or schedule it for later. To start a Hangout:

- Log into your Google+ account and hover your mouse over the left-hand side of your screen. A menu will pop up. Choose Hangouts.
- To start a Hangout immediately, you have two ways to do it:

  1. Choose Video Hangouts at the top of the screen and then choose Start a Video Hangout or
  2. On the bottom right-hand corner, choose Start a Video Hangout.

- To schedule a Hangout for later, choose Start Hangout on Air at the top of the screen.

- Install the plug-in at the prompt.
- Click Invite People and either send the participants the link to join the meeting or enter their e-mail addresses to have a link sent to them.
- Click "Start Broadcast" to the meeting.
- Hangouts on Air will automatically be broadcast to YouTube. Video Hangouts will not.
- Explore the apps on the left side of the screen. Some of the more popular apps are YouTube, Screenshare, and Chat.
- Make sure you click the phone at the top of the screen to end your meeting.

The Google+ support page has useful tips on starting and hosting Hangouts. In addition, there are people within the Google+ community called Hangout Helpers. These people host Beginner Hangout Saturdays, which are Google Hangouts you can attend to help you become more adept at scheduling and hosting Hangouts (Google 2014). Another way to get answers is to post questions within the Google Hangouts community, where other users will answer them. You can access the communities by using the same drop-down menu on the left-hand side of your screen that you used to access Hangouts. Customer service such as Hangout Helpers and the Google+ users themselves might explain why in a recent poll, Google+ ranked higher than Facebook in customer satisfaction. This also might be due to concerns about Facebook's privacy and security issues (Barlas 2012).

Now let's look at a real-life example of using Google+ for researching. In 2013 a study was performed that involved nursing students in a public health nursing practice course. Each student used Google+ and a Tablet PC. They also were required to use group investigation. This means students could choose their own research topics and research those topics themselves instead of having the instructor provide all of the information for them. Students were able to use other Google applications, such as Google Drive, YouTube, and Gmail by signing into their individual accounts (Wu et al. 2013).

Group investigation, an active and collaborative learning process, allows students to cater their learning activities to their individual learning styles and information needs. Because each learner is autonomous, they need good organizational and communication skills, thus adding to their ability to work collaboratively online. The nursing educator then becomes a facilitator during class and checks in on each group to see how they're doing. She might provide guidance on a concept or idea, encourage more participation from a particular student, observe how well the group works together, and ensure that students are active learners (Wu et al. 2013).

Here are the steps students followed to research a topic and create a group presentation (Wu et al. 2013):

- Introduce topics: Students are given guidelines for procedures. They are also told about rewards that will be given out to groups, which is done to motivate the students to become active learners. The instructor can then use Google Docs to upload supporting materials to Google+ for students to reference, and students can use their tablets and start virtual discussions.
- Form groups and select subtopics: The instructor and students both use circles to create groups to exchange information. Students then decide on a topic and what information they want to share within their circle. They will need to collaborate to successfully complete this step.
- Exploratory activities: Students divide up work and start sharing ideas. They also start posting research and deciding how to best store that research (i.e., does it all live on Google Drive?). This teamwork can encourage students to ask questions and depend on each other for assistance or clarification. They also will use their tablets to upload information, ask questions, or post photos or video to their circle. The tablets allow the students to participate anywhere.
- Prepare for presentation: Students can use Google+ to evaluate their research and decide how they present their information. The nurse educator will also help facilitate this step by providing guidance on methods and encouraging students to explore ways to create unique presentations.
- Presentation: Students do presentations and observe other presentations. This is how they learn from their peers. Each presentation is uploaded to Google+ for everyone to discuss and give feedback. This step helps students with both critical learning and academic discussions.
- Learners and educator evaluation: Students perform peer and group evaluations, and the educator evaluates the students' group investigation skills. The group with the highest score is rewarded, and their presentation will be published on Google+ so future students have an example of an exemplary group presentation.

Before you begin any social media venture, check with your organization's administration, compliance department, or management. In some instances, social media is restricted by law. For example, the financial industry's compliance rules are governed by the Financial Industry Regulatory Authority (FINRA) and the Securities and Exchange Commission (SEC). Companies in this industry must vigilantly monitor any social media information they release. This also includes whatever employees post about the company. How serious is this industry about compliance? The CEO of a publicly traded company recently tweeted that his company had a successful board meeting. He said numbers were good, and the board was pleased. He was fired within twenty-four hours of posting the tweet (Crosman 2013).

REFERENCES

Agence France-Presse. 2013. "Google Plus Racks Up Followers, But Not All Are Devoted." *Times of Oman*, May 1. www.rawstory.com/rs/2013/05/01/google-plus-racks-up-followers-but-not-all-are-devoted/ .

Balubaid, Mohammed A. 2013. "Using Web 2.0 Technology to Enhance Knowledge Sharing in an Academic Department." *Procedia: Social and Behavioral Sciences* 102: 406–20. doi:10.1016/j.sbspro.2013.10.756.

Barlas, Pete. 2012. "Score One for Google+ vs. Facebook." *Investor's Business Daily*, July 16. http://libproxy.csun.edu/login?url=http://search.proquest.com/docview/1026593207?accountid=7285 .

Barnes, Lescault, and Stephanie Wright. 2014. "2013 Fortune 500 Are Bullish on Social Media: Big Companies Get Excited about Google+, Instagram, Foursquare and Pinterest." *Charlton College of Business Center for Marketing Research, University of Massachusetts Dartmouth.* www.umassd.edu/cmr/socialmediaresearch/ 2013fortune500/ .

Benady, David. 2014. "Google+: Not a Social Network but a Social Layer." *Guardian*, February 5. www.theguardian.com/technology/2014/feb/05/google-plus-not-a-social-network-but-a-social-layer .

Chen, Sharon. 2013. "Meeting Them More Than Halfway." *Quirk's Marketing Research Media* (August 2013): 53. http://bluetoad.com/publication/?i=169141&p=52 .

Crosman, Penny. 2013. "How Software Can Help Banks Meet Social Media Rules." *American Banker*, March 27. www.americanbanker.com/issues/178_60/how-software-can-help-banks-meet-social-media-rules-1057873-1.html?zkPrintable=1&nopagination=1 .

DeLoma, Jamie. 2012. "Be Active in Your (Digital) Community." *Quill* (January/February): 37. http://digitaleditions.walsworthprintgroup.com/article/DIGITAL+MEDIA/961608/98521/article.html .

Erkoller, Alpetekin, and BJ Oberer. 2013. "Putting Google+ to the Test: Assessing Outcomes for Student Collaboration, Engagement and Success in Higher Education." *Procedia—Social and Behavioral Sciences*, 83(2013):185–89.

"50 Great Google+ Tips for School Librarians." 2012. *Online College.* www.onlinecollege.org/50-great-google+-tips-for-school-librarians/ .

Fredrick, Kathy. 2013. "Google . . . Plus." *School Library Monthly* 29(6): 23–25.

Google. 2014. "Start or Schedule Your Hangout on Air." Accessed June 9, 2014. https://support.google.com/plus/answer/4386744?hl=en .

Hines, Kristi. 2012. "How to Use Google+ for Business and Professional Branding." *Wordtracker.* Accessed June 10, 2014. www.wordtracker.com/attachments/google-plus-final.pdf .

Jain, Priti. 2014. "Application of Social Media in Marketing Library and Information Services: A Global Perspective." *International Journal of Academic Research and Reflection* 2(2): 62–75. www.idpublications.org/wp-content/uploads/2014/01/ .

Roscarla, Tanya. 2012. "How Google Plus Could Connect Students." *Tribune Business News*, December 17. http://libproxy.csun.edu/login?url=http://search.proquest.com/docview/1242099284?accountid=7285 .

Schultz, Don E., and James Peltier. 2013. "Social Media's Slippery Slope: Challenges, Opportunities and Future Research Directions." *Journal of Research in Interactive Marketing* 7(2): 86–99. doi: 10.1108/JRIM-12-2012-0054.

Stevens, Vance. 2013. "Tweaking Technology: How Communities Meet Online Using Google+ Hangouts on Air with Unlimited Participants." *Teaching Education as a Second or Foreign Language* 17(3): 1–15. http://tesl-ej.org/pdf/ej67/int.pdf.

Wu, Ting-Ting, Shu-Hsien Huang, Mei-Yi Chung, and Yueh-Min Huang. 2013. "Group Investigation Learning with Google Plus for Public Health Nursing Practice Course." *2013 IEEE 13th International Conference on Advanced Learning Techniques* (July 13–18): 238–42. doi: 10.1109/ICALT.2013.74.

Part IV

Searching

Chapter Twenty-Two

Advanced Search Strategies for Google

Teresa U. Berry

With its remarkable ability to pinpoint information from billions of web pages within a fraction of a second, Google has become the search engine of choice. It has become so entrenched in everyday life that in 2006 the verb *Google* was added to the *Oxford English Dictionary*. Given the ease of using Google, perhaps it is not surprising that, as McDonnell and Shiri (2011) point out, "users do not employ advanced search techniques or make full use of the features available to them in search engines" (15). While Google can usually anticipate what a user is searching for, there are times when someone spends hours futilely going through pages of unsatisfactory results. Using Google's advanced search features can help users by pushing more relevant links toward the top of the results list.

Google uses a program, known generically as a bot or web crawler, to browse the web by following the links it discovers on a web page. The Googlebot follows an algorithm, a set of step-by-step instructions, to determine which websites to crawl, how often to crawl them, and how many pages from each site to crawl. Google also uses sitemaps, a list of web pages provided by webmasters, to augment its crawling process. The crawling frequency of each site varies. News sites and other popular websites with constantly changing content are crawled several times a day, whereas sites with more stable content are crawled less frequently. Google makes copies of all the web pages it encounters and compiles them into an immense database. This database is what users are searching whenever they are Googling.

It is also important to understand what Google is not crawling. While Google can see the content of many file types (e.g., Adobe PDF documents and Microsoft Office files), there are some web pages it cannot process. For example, multimedia files, video content, and images serving as text are problematic for crawlers. Google may also miss dynamic pages, in which content is created on the fly every time the page is loaded. Subscription-based content or websites with registration requirements are sometimes inaccessible to web crawlers. Also, webmasters have the option to place instructions in a robots.txt file in their websites to ask bots not to crawl the site or to limit crawling to specific areas of the site. Furthermore, there are websites that Google deliberately does not index because their webmasters are using deceptive techniques to manipulate search engine rankings or are involved in unscrupulous web practices, such as spamming. Google also removes content in response to requests from government agencies and courts as well as from intellectual property owners and organizations that claim copyright infringement. As you can see, not all the web is discoverable through Google, and it can be advantageous to use other search engines and search strategies.

Google's success at returning highly relevant results is driven by PageRank, developed by Larry Page and Sergey Brin, the founders of Google. PageRank is based on the idea that a page's importance is based on the number and quality of the pages that link to it. Along with PageRank, Google looks at hundreds of factors, such as frequency and location of the keywords on a page, the content length of a page, and reputation of the website. Google (2011) revealed that it makes over five hundred changes each year as part of its ongoing effort to improve the search experience. Google also customizes your results by using personal information, such as your location, search history, pages you visited, and what ads you clicked. As a result of these dynamic elements, users will sometimes notice they are getting slightly different search results than they did before.

As you may have surmised, web design is an important factor that affects the effectiveness of search engines. As Google changes its indexing and ranking algorithms, website owners are concerned about where their pages appear in the search results. Many webmasters, particularly those associated with businesses, use search engine optimization (SEO) techniques to design sites that are crawler friendly and that perform well in the rankings. Unfortunately, some website owners do not take the same care in designing their sites. For example, a small local government agency may not care about Google's algorithms; their only goal is to publish the information on the web. For example, Google may not be able to find a local ordinance about garage sales that is buried in a large PDF file with a meaningless title. Your best bet may be to use Google to find the local government site, and then examine the site for possible resources.

THINKING ABOUT SEARCH

Now that you have some understanding of Google's content, let's consider how to search it. One of the most fundamental points to remember is that

Google indexes every word on the web pages it crawls. It does not ignore articles, prepositions, conjunctions, and other insignificant words that are often not searchable in traditional bibliographic databases. Using Russell and Bergson-Michelson's (2011) example, searching for *who* (World Health Organization), *the who* (the musical group), or *a who* (a Dr. Seuss character) shows that a minor word change can yield very different results. Likewise, word order is important because it is a primary factor in ranking search results. Searching for *table lamp* will find small lamps to purchase, but *lamp table* will find tables. And while searching for *hilton paris* does not completely eliminate *paris hilton* from the results, at least the hotel-related links are ranked higher than the pages about the celebrity. When the results are unsatisfactory, reconsider the search terms being used and their order.

Finally, take time to imagine the answer. Forming a mental picture of the page you hope to find will help you formulate a search strategy. Suppose you want to find information about women's rights in other countries. Let's think about how the answer might look:

- Who might have the information? United Nations, Amnesty International, women's rights groups, think tanks
- What words would I expect to find? Women's rights, human rights, world, global, gender equality, reproductive rights
- What format do I expect? PDF documents of reports, spreadsheets with statistics, web pages with articles, scholarly books, presentation slides
- What countries should I look at? Afghanistan, Pakistan, Middle East, Africa, India

These examples illustrate the different directions a search can take. After considering these possibilities, you decide to focus on government documents that discuss women in Afghanistan or Pakistan. Here is one approach to searching Google:

women OR human OR gender rights afghanistan OR pakistan site:gov file type:pdf

This search strategy yields very different results than simply entering *women's rights pakistan*. Now let's look at some of Google's advanced search techniques.

IMPROVING SEARCH RESULTS

You can fine-tune results by applying the filters under the Search Tools menu at the top of the results page. The options vary according to the type of search you are doing (e.g., web, image, video, etc.). For example, a search for cooking ingredients (*chicken mushroom rice*), will have recipes in the results, but clicking on Search Tools will give you options to narrow results by ingredient, cooking time, and calories. Google also offers a set of specialized search operators that can be entered directly into the search box. Alternatively, the same features are available on Google's Advanced Search page located under the Settings link. These techniques are primarily a way to influence the results ranking so that the desired information appears higher in the list. Keep in mind that Google is always observing search behavior to better understand what users are asking and to predict what they want to find. Due to its ever-changing algorithm, search features and operators can disappear, as in the case of the synonym (~) operator that was dropped in 2013. Given that caveat, here are some advanced search techniques to try.

Time Period

One useful filter is by time, which allows you to limit to the past hour, day, week, month, year, or a custom date range. This option is particularly helpful if you want to find the latest content (*walking dead* within the past week) or older information (*health care reform* during the 1990s). Emphasize to your patrons that time reflects the date of the web page, not necessarily the time period covered by the content. Searching for web content about the Boston Tea Party during the 1770s will be fruitless. However, changing the search from Web to Books will reveal a number of primary resources written in that time period. It is important to realize that a computer program is determining the page's date, so always verify the date by examining the web page itself.

Exact Words and Phrases (*" ", intext:*)

Since Google automatically incorporates synonyms and word variations into the search, use double quotes to search for an exact word or phrase. Alternatively, you can filter your results using the verbatim filter found under Search Tools. Google tells you ("Showing results for") if it used alternative words in the search, a practice that usually happens with misspellings. Using double quotes will help suppress Google from including synonyms, word variations, corrected spelling, and personalized results based on your search history. For example, *"john stewart"* helps minimize the number of results about *The Daily Show* host Jon Stewart. You can also take it a step further and search for exact words in exact order (*" "dogs" "cats" "birds" "*).

Another effective method for searching exact words and phrases is to use the *intext:* search operator. Try *intext:majefty* to find examples of early English texts (computer programs sometimes interpret the letter *s* as an *f*). Whereas the double quotes/verbatim option will look for the search terms in the URL, title, and other parts of the web page, *intext:* will force Google to look only in the main body of the page. Use *intext:* for single words and phrases in quotes (*intext: "winter of 2014"*). Use *allintext:* for multiple words (*allintext: kennedy obama clinton*).

Including Alternative Words (*OR*)

Although Google automatically includes synonyms in its search as indicated by the bolded words on the results page, you may notice that some synonyms were omitted. You can then use *OR* to ensure Google includes the desired term (*women's rights world OR global*). Note that *OR* must be in capital letters, or it will become a search term. Also, OR is meant to be used to link single words (*world OR global*). For phrases, use double quotes (*"women's rights" OR "human rights"*).

Excluding Words (-)

Sometimes Google's ranking algorithm returns results cluttered with multiple meanings or popular culture phenomena. To help filter results, try using the minus sign (-) to exclude words (*big bang theory -cbs*). Due to Google's predilection to expand your search with synonyms and word variants, you may end up with a long string of excluded words. In those cases, consider adding terms to focus the search. (*big bang theory universe*).

Specifying a Website (*site:, inurl:*)

The *site:* operator allows you to search within a specific website or domain. This strategy is particularly useful when navigating large, complex websites (*health care site:gov*), looking for a certain viewpoint (*boston marathon site:bostonglobe.com*), or excluding websites from results (*wikipedia -site:wikipedia.org*). The drawback to restricting to a single domain is the possibility of missing relevant materials. For example, the U.S. National Archives and Records Administration has partnered with Fold3.com to digitize military records, so those documents will appear in a .com rather than a .gov domain.

It is also possible to search for words appearing in the URL. Whereas the operator *site:* restricts the search to a specific URL, *inurl:* searches multiple sites and domains. For instance, *site:census.gov* limits the search to the U.S. Census Bureau's website; *inurl:census* will search websites belonging to state government agencies, organizations, *Wikipedia*, and more. The operator *inurl:* works for single words and phrases (*inurl:"katrina recovery"*). Use *allinurl:* to search for multiple words (*allinurl: michigan census)*. Of course, you are at the mercy of whoever decides the names of the links.

File Types (*filetype:*)

When choosing a search strategy, it helps to predict whether the desired information will exist in a spreadsheet, PowerPoint, Word document, or some other format. The *filetype:* operator will limit results to files with the designated extension (e.g., pdf, xls, doc, or ppt). For example, to find some training materials for a CPR class, try searching *cpr guide filetype:pdf.* Looking for statistics on the gross domestic product of Latin American countries? Try searching for *gross domestic product latin america filetype:xls* to find spreadsheets containing statistical data. If you do not have the software application needed to open a particular file, the text will still be readable by clicking on the green arrow next to the URL and accessing the cached version of the page.

Wildcards (*)

Like many traditional databases, Google allows wildcards using the asterisk symbol (*). Since Google already looks for word variations, the asterisk represents a single word (*"to * or not to *"*) and does not function as a wildcard for word endings. You can use wildcards in a fill-in-the-blank approach to finding information (*senator * voted no on gun control*). It is possible to use more than one asterisk. For example, *allinurl: katrina * * recovery* will retrieve sites with "Katrina Gulf Coast recovery" in the URL.

GOOGLE BOOKS AND GOOGLE SCHOLAR

Google offers two specialized search engines that are worth noting. Google Books provides full-text searching of books provided by participating publishers or library digitization projects. Users appreciate the ability to see the content of rare books, scholarly and professional titles, textbooks, fiction, children's books, and much more. Students can go beyond *Wikipedia* and use Google Books to find background information on a topic by choosing keywords like *introduction*, *textbook*, or *guide*. Google Scholar searches material available from academic publishers, universities, United States courts, and other scholarly resources. Because these search engines are searching the full text of books and journals, Google Books and Google Scholar are particularly valuable to researchers who do not have access to commercial bibliographic databases. Be aware that it may take some time for Google Scholar to crawl these websites, so materials will appear more quickly in Google's web search. Both of these search engines are popular with users who struggle with traditional bibliographic databases.

KNOWLEDGE GRAPH

Introduced in 2012, Knowledge Graph refers to the information boxes that appear on the search results page. According to Singhal (2012), the Knowledge Graph contains facts about and relationships between people, places, and things and presents information based on what users have searched in Google. Users no longer have to click a link to get quick answers to their questions because the information is strategically positioned on the results page. For example, search *chips vs fries*, and you will get a table showing a side-by-side comparison of their nutritional values. Knowledge Graph's information comes from various public sources, some of which are authoritative (United States government agencies, World Bank, etc.), but the content appears to come mostly from *Wikipedia* and Freebase.com, another volunteer-based knowledge repository. Unsurprisingly, Knowledge Graph has errors, so Google provides a way to report problems. Sullivan (2012) reports, "Google will use a combination of computer algorithms and human review to decide if a particular fact should be corrected." The convenience and visual appeal of these displays make them hard to ignore; however, users are urged to evaluate and verify the information.

THE FUTURE OF SEARCH

Although Google has been the dominant search engine for many years, it is facing challenging times. Duggan and Smith (2013) found that 21 percent of cell phone owners, which includes those who do not own a computer, use their cell phone as their primary device to access the Internet. Domain-specific apps, such as Twitter and Facebook, create silos and take eyes away from Google and its primary source of revenue—search ads. With smartphone owners spending only 20 percent of their time in web browsers, Winkler (2014) suggests that search engines are becoming less relevant. Web crawlers have difficulty seeing content in mobile apps, and Google needs to develop partnerships with mobile app developers in order to crawl the apps and deliver mobile ads. Google has already taken steps to meet these challenges. It has begun to mirror mobile interfaces by introducing new features, such as voice-activated search and a flights booking site. As the data in Knowledge Graph grows, web traffic to competing sites may decline as Google becomes the destination, not the path to another's website. Given Google's history of success and continuing expansion, it will be interesting to see how searching the web will change.

SUGGESTED RESOURCES

The main challenge in becoming an expert searcher is keeping up with the constant changes to Google. While there are many opinions from search engine experts about Google search, the most reliable ones are the comments made by Google employees themselves. These tidbits can be found in news stories, Google product forums, social media sites, and comments sections of these websites. Below are a few suggested resources for learning more about Google search:

Inside Search (http://www.google.com/insidesearch/): Provides information on how search works, tips and tricks, results ranking, and more. Be sure to check out the blog for updates on new features and changes to the search algorithms.

Search Education (http://www.google.com/insidesearch/searcheducation/): Lesson plans and activities for teachers, self-paced courses on power searching, and webinars on search skills and information literacy.

Search Engine Land's Guide to Google (http://searchengineland.com/guide/google): A source of news and information about Google and all its products.

SearchReSearch (http://searchresearch1.blogspot.com/): A blog by Daniel Russell, a Google research scientist, who gives tips about searching Google along with weekly search challenges.

Webmaster Tools (https://www.google.com/webmasters/): Although the information is geared for website designers, it provides insight into how Google indexes and ranks content.

REFERENCES

Duggan, Maeve, and Aaron Smith. 2013. "Cell Internet Use 2013." Washington, DC: Pew Research Center. http://www.pewinternet.org/Reports/2013/Cell-Internet.aspx .

Google. 2011. "How Google Makes Improvements to Its Search Algorithm." YouTube video, 3:53. August 24, http://youtu.be/J5RZOU6vK4Q.

McDonnell, Michael, and Ali Shiri. 2011. "Social Search: A Taxonomy of, and a User-Centred Approach, to Social Web Search." *Program: Electronic Library and Information Systems* 45(1): 6–28. doi:10.1108/00330331111107376.

Russell, Daniel, and Natasha Bergson-Michelson. 2011. "Even Better Search Results: Getting to Know Google Search for Education." YouTube video, 59:46, October 7, http://youtu.be/f2jqwNxq1cM.

Singhal, Amit. 2012. "Introducing the Knowledge Graph: Things, not Strings." *Official Google Blog* (blog), May 16. http://googleblog.blogspot.com/2012/05/introducing-knowledge-graph-things-not.html.

Sullivan, Danny. 2012. "Google Launches Knowledge Graph to Provide Answers, Not Just Links." *Search Engine Land* (blog), May 16. http://searchengineland.com/google-launches-knowledge-graph-121585.

Winkler, Rolfe. 2014. "Google Searches for Role in App Age." *Wall Street Journal*, March 10.

Chapter Twenty-Three

Evaluating the Sources of Search Results

Jennifer Evans

Students, patrons, and, yes, even librarians sometimes turn to Google for their research. The Internet is a vast store of knowledge, but not all the information on each website is created equally. The searcher must be able to distinguish accurate information from inaccurate. This process is called website evaluation.

The user must consider:

- the author and his/her authority
- the author's purpose and objectivity
- the information's accuracy
- the content of the site
- the author or publisher's reliability and credibility
- the currency of the information
- and any links found on the site

In this chapter, you will find a compiled list of questions, mostly courtesy of the website of the Georgetown University Library ("Evaluating Internet Resources" 2014) that you can ask about each website to determine the scholarly value of the information presented. However, some feel this traditional method of analyzing website content is incomplete. This chapter includes these arguments as well as the suggestions for change. Whichever approach you choose to use, you will be prepared to evaluate websites found during Google searches.

WHY BOTHER?

You may wonder why you need to bother with evaluating websites. The answer is that anyone, anywhere, with any motive, can put whatever information he or she wants on the Internet. Whether you need a simple health question answered or are doing a doctoral dissertation, you need to be certain of the authority and the accuracy of the information you find on the Internet. If you are searching for information on cancer, you will find a number of websites of universities, hospitals, nonprofit cancer organizations, individuals with the disease, pharmaceutical companies, and companies selling alternative treatments. If you are writing a research paper, you risk failure and embarrassment if you use faulty information as the basis or support for your thesis. If you are researching treatments, the wrong information could lead to physical harm. Either way, sifting through your search results for factual information is crucial.

AUTHORSHIP AND AUTHORITY

The author and his or her authority should be the first things you look for when evaluating information found on the web. You need to be able to ascertain (1) that the author is identified, (2) that the source of the information is a knowledgeable professional in the field, and (3) that the author and/or publisher have a reputation for adding to the scholarly field rather than publishing for personal or commercial gain. If the author/creator of the content is not identified, that should be an immediate red flag. Note: Someone with the title "webmaster" or something similar is not necessarily the author. Often, a web designer or other technology expert will manage the site, but the content is written by another person. If you cannot verify the source of the information, you may not want to include it in your findings. If an author is listed, you can find information about him/her by looking at the author's home page or doing a search on the Internet (for tips on doing an effective Google search on an individual, see below). You can ask yourself the following questions about the author to help determine the authority of the page.

- Is the author identified on the page?
- Is the author's contact information available on the page?
- What are the author's credentials?
- Is this person an educated professional in the field or a layperson?
    - Is the author representing a group or organization? What are the organization's credentials?
    - Is it a university? A hospital? A nonprofit group? A political group?
- What information can you find on the web about the author?

Here is a trick to find the most complete set of results using Google ("Evaluating Web Pages" 2012):

1. Search the author's name without quotes (e.g., *John Doe*)
2. Search the author's name enclosed in quotes as a phrase (e.g., *"John Doe"*)
3. Search the author's name enclosed in quotes with an asterisk (*) between the first name and the last name (e.g., *"John * Doe"*). The asterisk can stand for any middle initial or name. Note: This only works on the Google search engine.

PURPOSE

Determining the purpose for which this information is provided is a next logical step in the website evaluation process. Why was this site published? It is best to use information that is provided for a scholarly purpose, such as to educate and inform. You may come across websites during your research in which the author's purpose is to persuade the audience to one side of an issue or even to sell a product. The authors or publishers of these sites are biased and have an agenda. Ask yourself: What is the purpose of the author and the site?

- To educate and inform?
- To entertain?
- To persuade?
- To sell a product?

And who is the intended audience?

- Scholarly audience or experts?
- General public or novices?

OBJECTIVITY

When evaluating a website, it is important to consider the author's and/or publisher's objectivity. The information should be free of bias. Information presented should be fact rather than opinion. Information from websites by authors who attempt to sway your opinion should be used with care. The following questions will help you to determine the author/publisher's objectivity:

- Is the author or publisher biased?

- Is the information presented fact or opinion?
- Is the author or publisher attempting to sway your opinion to one side of an argument?
- Is the author affiliated with an organization?
- Does this affiliation seem to bias the information?
- Does the author have the official approval of the organization?

ACCURACY

When you have determined that the information is free of bias, you must ensure that the information itself is accurate. Any claims, facts, statements, or statistics presented need to be substantiated. If the author performed original research, he/she should provide the research methods used. Ensure these methods are sound and thorough. If the author used information from other sources, you need to examine these sources to make certain they are trustworthy. Use caution, especially if the information presented is contradicted by other reputable sources. Use the following questions to help you verify the accuracy of the information given.

Did the author do original research to determine the findings?

- If so, did the author use proper research methods?
- If not, from where did this information come?
- Is the author using information from another source in the spirit in which it was intended? In other words, did the author selectively pick and choose content or use a statement out of context to suit his or her own needs?
- Are the facts presented accurate?
-
    1. How do you know?
    2. Can you verify the facts with evidence or by using other sources?

- Is the page free of errors (misspelling, typographical errors, etc.)?

CONTENT

Once you have determined the authority and objectivity of the author and the accuracy of the information presented, the next step is to analyze the gist of the content of the page. Many pages appear to be legitimate, complete with references and statistics, but even these may not be credible. For example, an online and print publication called *The Onion* presents its material in the form of a news story, but all of the information is satirical in nature and fictitious and is intended for entertainment purposes only. Also, you should exercise caution when using the website *Wikipedia*. While some information may be accurate, anyone is capable of putting any information on the site. Most teachers and professors will not accept any references from *Wikipedia* in a research paper. Furthermore, even serious-minded sites can have false or exaggerated information, even fake references. These questions can help you decide whether or not the content of a website is trustworthy:

- What do other people say about the author or the site's content ("Evaluating Web Pages" 2012)? Do a Google search and find out.
- Is the page written to be ironic, humorous, or satirical ("Evaluating Web Pages" 2012)?
- Are there references or sources listed for the information on the site? Are they real? Do they contain factual information?
- Does the information seem realistic, or is it exaggerated?

RELIABILITY AND CREDIBILITY

Reliability and credibility are closely related to authority and accuracy. Basically, these factors can help you to determine whether or not the information presented on a website is trustworthy. You examine the source of the information to decide if you can depend on it to present factual data. One way to do this is by taking note of the domain of the website (e.g., .gov, .com, .edu, etc.). Perhaps facts from an educational institution may prove to have more merit than those from a business. Here is a list of website domains and the type of organization that is assigned to use each:

.com: commercial/business
.org: organization (Note: The domain ".org" was once exclusively for use by nonprofit organizations, but that is no longer true.)
.net: network provider
.edu: educational institution
.gov: government
.mil: military
~: Presence of a tilde sign usually means that the page is run by an individual rather than an organization, institution, or business

Furthermore, you should use the same criteria to evaluate websites that you are using to evaluate other sources ("Evaluating Web Pages" 2012). For example, you want to use an article you found in a monthly print publication. You need to determine whether it is a scholarly journal or a popular magazine. Anything scholarly, especially if it is a peer-reviewed publication, is probably trustworthy whereas many (but not all) popular magazines are generally published for entertainment purposes. The same can be said for websites.

Are you using the same criteria to evaluate web resources as you are other sources, such as print publications or databases?

CURRENCY

Sometimes the currency of information can be important. For example, medical or scientific data can quickly become outdated since new research results come out so often. However, in some cases, currency is fairly irrelevant. For instance, when doing research on papers written about Jane Austen novels, currency does not really matter because her novels were written so long ago. If you find yourself in need of the most up-to-date information, ask the following questions when evaluating a website:

- When was the site first published?
- When was the content of this website written?
- When was the last time this site was updated?
- Does the content seem outdated compared to other information sources?

One method you can use to restrict your search to the most current material is by doing an advanced search in Google (Bromann-Bender 2013). Simply click Settings and then Advanced Search. Then you can narrow your search results by selecting a time frame from the "last update" menu. Note: The default setting is "anytime."

LINKS

Oftentimes, credible websites have links to outside sources since the authors' purpose is to educate. They want you to have access to as much factual information as they can provide. The fact that links are present does not necessarily mean a website has accurate information, but it is one of many things to consider when evaluating websites. You should examine any links that are present—they can give you clues as to whether or not the website is trustworthy. Use the following questions about links to help you determine the integrity of each website:

- Are there links to outside sources?
- Are the outside sources relevant?
- Do the outside sources seem legitimate and accurate?
- What is the purpose of the linked sites?
- Do the links still work, or are they dead or broken?

RESEARCH READY, A WEBSITE EVALUATOR

Research Ready is a paid service that provides resources on doing research for educators, administrators, and students. The website offers a free Website Evaluator (2014). Simply type in the URL of the site you wish to evaluate, and you will be taken to that website with a toolbar from Research Ready. The toolbar will walk you through evaluating the website by presenting questions you should ask regarding the purpose, accuracy, and authority of the author and the publisher and the relevance and currency of the content, and a final review in which you determine whether the site is credible or not. Note: the Website Evaluator does not make the decision of whether or not a site is credible—it helps the user to make that decision on his or her own. While not exhaustive, this free resource provides some practice for evaluating websites on your own.

BEYOND THE RUBRIC: NEW THOUGHTS ON WEBSITE EVALUATION

Some scholars see the aforementioned method of evaluating websites as inadequate or even inaccurate (Ostenson 2014). The main argument for this viewpoint is that it focuses on a simple question-and-answer assessment method rather than engaging critical thinking from users. For example, this method can create false positive results (Ostenson 2014). Just because all of the predetermined pertinent information seems to be available does not mean that the information provided is accurate. On the other hand, the presence of dead links or the fact that the page has not been updated recently does not necessarily mean that the information is wrong. Ostenson (2014) suggests a three-pronged strategic approach to evaluating information found on the web: contextualizing, sourcing, and corroborating.

Contextualizing

Contextualizing involves performing initial research from some trusted source, such as a scholarly database, to gain some background information on your topic that is reliable. What you learn from this preliminary inquiry can help you put information you find on the Internet into context and will also help you find keywords to use in your Google search (Ostenson 2014).

Sourcing

After you have done some introductory research, you are ready to do a Google search for information. Once you have found some information, you must source it. Sourcing involves evaluating the source of the information you find on the Internet, much in the same way as the previously discussed traditional question-and-answer model. Ostenson (2014) advises that you should seek to determine the following:

- Who wrote this information, and what are his/her qualifications?
- What type of document is this (news article, blog, academic paper, etc.)? Is this appropriate for your research? How does the type of source influence the credibility of the information?
- Why is the author publishing this information? What is his/her agenda, if any?

Ostenson (2014) also suggests looking at the authority, accuracy, objectivity, currency/timeliness, and coverage of the information presented to verify credibility.

Corroborating

Once you have evaluated the source of the information from your Google search, you should corroborate the facts presented with those of other reliable sources (Ostenson 2014). Data and ideas that are supported from multiple sources are likely to be dependable. If you find conflicting information, go back to the sourcing step to determine which resource is more trustworthy. If the source of the information is questionable and the information is contradictory to that of other credible sources, you should question its reliability. Keep in mind, though, that controversial topics can be complicated, and some disagreement among experts should be expected.

CONCLUSION

Websites can be a great source of information when doing research, but prior to using this information, you need to evaluate the website, including the source of the information presented as well as the information itself. In fact, prior to doing a Google search, you may want to perform some initial research from reputable sources, such as a database at a university, to obtain some factual information in order to put any information found on the Internet into context. Once you do this initial research, you are ready to do a Google search to find websites with pertinent information for your research. You need to evaluate each website prior to including its information in your findings. You will need to find out who the author is and determine his or her authority on the subject matter. You will need to establish the author and/or publisher's purpose in offering the information. Be wary if the purpose is to sway your opinion or to sell you something. Ensure the author's objectivity on the matter. Authors who are biased can, even unintentionally, frame information or data in such a way as to suit their own needs. You must make certain that the information presented is accurate. You must then examine the content of the site to check that all data is factual and is not exaggerated or ironic, humorous, or satirical. Check the date the site was published and the date it was last updated—some findings or theories can quickly become outdated. If links are present, make sure that they still work and that they link to reliable sites. After completing these steps, sometimes called "sourcing" (Ostenson 2014), you are ready to corroborate your online findings. This involves checking facts and statements you have found on the Internet with those found in other reputable sources. If a claim found on the Internet conflicts with those found in more trustworthy sources, such as peer-reviewed journals, you should consider any scholarly report as the more credible source.

Anyone can put anything on the Internet—it is not regulated. When performing research, you want to get the most accurate information possible. Evaluate each website from which you want to use information to ensure that you are getting the most reliable data. A Google search can lead you to the most up-to-date facts because there is less lag time as compared to the print publication process. Use this source wisely by applying these website evaluation techniques.

REFERENCES

Bromann-Bender, Jennifer. 2013. "You Can't Fool Me: Website Evaluation." *Library Media Connection* 31(5): 42–45.

"Evaluating Internet Resources." 2014. Georgetown University Library Research Guides. Accessed August 27, 2014. www.library.georgetown.edu/tutorials/research-guides/evaluating-internet-content.

"Evaluating Web Pages: Techniques to Apply and Questions to Ask." 2012. University of California Berkeley Library. Accessed August 27, 2014, www.lib.berkeley.edu/TeachingLib/Guides/Internet/Evaluate.html.

Ostenson, Jonathan. 2014. "Reconsidering the Checklist in Teaching Internet Source Evaluation." *Libraries and the Academy* 14(1): 33–50.

"Website Evaluator." 2014. Research Ready. Accessed August 27, 2014. http://webeval.researchready.com.

Chapter Twenty-Four

Free, Easy, and Online with Google Sites

John C. Gottfried

WHAT IS GOOGLE SITES?

Google Sites is a free web service offered by Google since spring 2008. Before this they offered a web page service called Google Page Creator, which was discontinued in favor of the more flexible current form. Google Sites may be used both as a wiki (in which groups can work and communicate interactively online), or as something more similar to a traditional website. Pages can be created from any of a group of web page templates serving a variety of specific purposes and tastes. There are a few limitations in layout and design flexibility, but Google Sites provides a user-friendly experience useful for anything from library staff intranets and guides to publicly accessible web pages for groups, classes, or subjects.

In addition to being free and very easy to use, Google Sites has several solid advantages for the user:

- Google Sites requires no programming or database management. It is, in fact, maintained entirely online, so it does not even require downloading any software to your computer.
- You can access, create, and manage your Google Sites account from any location.
- Your Google Sites account is independently managed: the user does not need to comply with any controls other than those imposed by Google itself. You also retain ownership of any content you create for the pages.
- As you may imagine, Google Sites integrates easily with other Google products and services.

- At this writing, the sites are ad-free.

And, while Google Sites does offer many benefits, the user should be aware of certain potential drawbacks as well:

- It does not have the design flexibility of a traditional website.
- It works best with Google Apps and other services, so it may not be compatible with many popular applications and widgets.
- While Google makes no claim on your content, it does own the site itself. You cannot transfer the site directly to another domain—you would have to recreate the pages in another system.
- By default Google Sites' URLs begin with the standard "sites.google.com/site/." Changing to another site domain is possible, but may be a more complicated process than many are willing to tackle.
- Remember that Google has dropped a number of popular applications in the past, sometimes without much warning. At their discretion, they could do the same with Google Sites.

Users should take all of Google Sites' pros and cons into careful consideration when making a decision as to whether it is the right tool for their purposes. On balance, however, Google Sites is a great free service, and it would be relatively easy to replace if necessary. It provides all of the basic services and flexibility needed to create effective basic web pages and guides. It permits collaboration on many levels—users can choose to share your site across the web, or with a select group of partners (each with specific levels of administrative access). The number of available templates is expanding, and it is relatively easy to create a custom template if desired. Finally, if Google Sites does not work well for a given individual or group, it costs no more than the limited amount of time it takes to set up a few experimental pages.

SETTING UP YOUR GOOGLE ACCOUNT

Setting up a Google Sites account is generally a quick and easy process. Like all Google products, Google Sites may be accessed from a Google account. If the user doesn't have a Google Account, he or she can find the website by—appropriately enough—doing a Google search. When you have found the Create an Account page it is very easy to fill out the form and start up your Google Account.

An important and potentially complicated consideration is deciding which type of Google account you wish to create. In broad terms there are currently two types of Google accounts: standard and Google Apps. Anyone can open a standard Google account. The standard account includes dozens of common Google applications, including Gmail, a suite of office applications, calendar, YouTube for videos, Picassa for pictures, and, of course, Google Sites. There is no charge whatsoever for a standard Google account.

Google Apps works a bit differently from the standard account. It is normally an institutional account, and Google Apps accounts are administered by the organization, whereas standard Google accounts are administered by Google. It allows the organization to better control the access and use of its institutional account. Google Apps provides access to many of the same applications as the standard account, but not all. Google Sites is one of the programs available in a Google Apps account, and the storage limits are considerably higher than for the standard accounts. At the time of this writing, Google Apps accounts are available free of charge to educational institutions and registered 501c3 nonprofit organizations, but there is a charge for businesses and government organizations (including public libraries). The exact conditions and fees vary by type and size of the organization, so check with Google for more information.

GETTING STARTED WITH GOOGLE SITES

Once you have access to a Google account, you are ready to access Google Sites at https://sites.google.com/. You can also find Google Sites with a simple Google search, or you can find it on the Google Products page under the Home & Business section. The home page is surprisingly simple: you will see a listing of the sites you have created (if any), and you will see a large red button with white lettering that says "Create." To create a web page, simply click the Create button.

The system as a whole is largely intuitive and very easy to use once you understand the basics, but there is a learning curve in the beginning of the process. If you are setting up the site yourself, Google makes the experience manageable for just about anyone. The Google Sites Support Center is here: https://support.google.com/sites/?hl=en. The help function on Google Sites is very effective, and the huge number of Google Sites users in the world means that you are never more than a quick Google search away from tips and advice to walk you through almost any problem.

The first decision you must make, for example, is the type of website you need. In some instances the answer may be obvious, but users unaccustomed to creating websites and pages may need a little coaching on the alternatives available. Google Sites provides useful templates for a number of purposes including pages for clubs, intranet sites for an organization, publicity sites for events, or collaborative websites for group projects. For the uninitiated, it is worth the time to create several test sites and try out some of the templates to get a better understanding of the range of possibilities. Any Google site can be deleted at any point, and you can adjust templates and other layout features at any time—so feel free to experiment and make mistakes.

If the goal is to create a website for a club, for example, Google Sites templates will provide you with a site design that will get you started quickly and easily. You can choose a design you like, then customize the page as desired. The template includes links to a number of mock-up pages such as club news, information about the club, a calendar of club events, and others. Each of the pages can be adjusted to the user's needs, and more pages and links can be added at will. Rather than a website, for instance, the user may wish to create an intranet site—a site designed specifically for members of a specific workgroup or organization. An intranet template might allow a group to store and access documents, hold online discussions, maintain a directory of members, and provide a calendar of events. Access can be limited to specific group members, offering a site that is both secure from outside access, yet highly accessible to the group.

Google Sites offers access to dozens, perhaps hundreds, of these ready-made templates, but users are also able to create their own sites from scratch. By starting with a blank template users are free to choose from among a large number of layout and design options. They can, for instance, decide whether they wish to add a header and footer on the page, and they can choose the number of columns and sidebars. They may wish to apply a preselected design theme (where colors and shapes of boxes are already determined) or they can personally pick out their own colors, type fonts, and other design elements. When you have achieved your ideal page design you can save it as a template, which can then be used again for future pages and projects. You may even wish to share your design with others, in which case you can save your template to the Google Sites template gallery.

ADDING CONTENT TO A SITE

Once the user has created the site it is time to add content. You can choose to add almost any type of content you wish, and there are often several very easy alternative methods of managing specific media. Below are some suggestions for some of the most common items people may wish to add to a site, but this list is by no means exhaustive.

- *Images:* You can add images directly to your pages, but in order to save memory on your site (important if you are working within the limits of a standard Google account), you can also link to the URL of a picture online, or you can upload from a Google Picasa photo account.

- *Videos:* You can provide access to videos through Google Sites in several ways. The simplest method is to simply link to a video on another website using the video's URL. You can also embed a video directly from Google Drive (Google's online cloud storage). YouTube is another easy alternative—because Google owns YouTube, it is very easy to mount YouTube videos through Google Sites using the Insert tab.
- *Maps*: Inserting a Google map in your site is a simple matter of using the Insert Map command and entering the address for the desired location. The Google map will be inserted in the page, and viewers can click on the embedded map to go the full Google Maps site.
- *Calendar*: Google Calendar is another easy option for your site. You create the calendar through your Google account—the calendar might include events, schedules, due dates, or other items as appropriate. In editing mode on the Google Sites page, you add the calendar from the Insert tab. You do have some control over some aspects of the embedded map (such as size and time period shown), but color and design options are preset.
- *Search Options*: Google supplies a search box on Google Sites pages. By default, this will search the site, but it can be changed to search multiple sites, an entire domain, or the entire web.
- *Other Widgets and Gadgets*: In addition to the standard options listed above, Google Sites offers access to a number of other gadgets, apps, and widgets. These may be accessed through the Insert tab and downloaded to your site. The choices range from practical applications like weather reports, clocks, and news feeds to oddities such as virtual pet rocks, UFO maps, and an online Magic 8-Ball.

The options above are among the most popular, but there are many other possibilities, and the list of options is constantly growing. Check the Google Sites help functions and user forums for tips and advice on customizing sites and content.

ORGANIZING YOUR SITE AND LINKING PAGES

Using the templates described above, the user can benefit from any number of basic but effective premade site designs. No matter how complete the template, however, most users will eventually need to add more pages to the site. Other users will wish to create page designs from scratch. Sooner or later, in other words, you will need to create one or more web pages, and you will need to make sure they link and display properly within your site.

To create a new page, go to your home page and click the New Page icon at the top of the screen (page symbol with a plus sign on it). You need to choose an appropriate name for your page—make the title descriptive, but keep it short and avoid symbols beyond basic letters and numerals. Next you will need to choose the type of page you wish to create. Google Sites supplies several options:

- *Web Page*—this is the most basic form. It is simply a blank page. It will automatically conform to any template or thematic designs you have designated for the site. All standard page inputs apply—you can control the page layout within normal parameters, and you can add tables, photos, videos, widgets, and all of the other items permitted for any Google Sites page. Text input is easy and straightforward using the basic word processing commands. You can attach documents at the bottom of the page and permit comments from readers.
- *Announcements*—this page helps you create a list of news items or announcements. It is easy to post an announcement by clicking the New Post button and typing or pasting your message. You can also add images, maps, calendars, and all the other options available in other Google Sites text boxes. Announcements will appear in the navigation menu as subpages of the primary Announcements page.
- *File Cabinet*—designed to hold most types of electronic files including documents, spreadsheets, audio files, web links, and so on. Files can then be sorted into folders.
- *List Page*—lists of various types can be entered in a row and column format that is easy to create and read. Items can be sorted by viewers according to any of the columns. Google Sites offers several premade list pages:

    1. Action Items—this list includes columns for the name of the person assigned to the task, a description of the item, status or resolution, and a yes/no completion checkbox. It allows the user to track separate tasks involved in a larger project.
    2. *Issue List*—this is designed to track problems or comments that arise in a project. There are columns for the name of the person registering the comment, the person responsible for resolving the issue, a priority designation (P1 through P4), a description of the item, and a description of the status or resolution.
    3. *Unit Status*—permits you to divide a project by units or features, tracking the status of the feature, assigning a person responsible, a description of the feature, and supplying a web URL to view a page about the specific feature.
    4. *Create Your Own*—if none of the predesigned list formats work for your project, you can create one from scratch. You can create columns to be filled out with text, date, a drop-down menu of choices,

or a URL to another page. By default items are sorted by date last edited, but this can be changed.

The page options listed above should cover most common situations. Remember, however, that you can create your own customized pages for other purposes. Once you have created your new web page, a link to the new page will automatically show up in your site's navigation menu. You can also choose to list your new page as a subpage of any of the other pages in your navigation menu, or you can link it to any of your other Google Sites. If you wish to delete a page, go to the page you want to delete, then to the Page Actions menu (look for the gear icon in the upper left corner of the page). You will see a command to delete the page. You can view and restore deleted pages for up to thirty days by clicking on Manage Sites in your Page Actions menu, then going to the Deleted Items link on the left sidebar.

SITE LAYOUT AND NAVIGATION

Another very important aspect of site management is to ensure that viewers can navigate the site easily and confidently. You can edit navigation menus and mapping by clicking on the Edit Site Layout link in the Page Actions menu (the gear icon in the upper right corner of the page). Here you will find several options for site navigation. You can use the standard sidebar navigation menu, and you can edit the sidebar in a number of ways—you can move it from right to left, and adjust the size. You can add recent activities, new links, recent page authors, text, images, and more. You can also provide navigation horizontally across the top of the page in the form of tabs, boxes, and links. The Edit Site Layout page is also where you can edit the footers and headers (the bands at the top and bottom of the pages).

GENERAL TIPS AND RECOMMENDATIONS:

- *Mobile Accessibility*—making your site accessible to mobile devices is easy on Google Sites: in the Page Actions menu go to the Manage Site link, then scroll down and click the appropriate box to allow your site to automatically adjust to mobile screens.
- *Keep Your Site Simple*—don't use more pages than you need on your site, keep the language simple, and make sure users can navigate to any page in your site from any other page in your site. Users will give up on a site that is too complicated to read or navigate.
- *Make Your Site Useful*—give your viewers content they want and need. Be particularly cautious of overloading the viewer with self-promotion—

viewers will visit your site in response to their own needs, not your ambitions or goals.

- *Stretch Your Memory*—if you do not have a Google Apps account and are building your Google sites through a standard Google account, you will not have much memory to work with (about 100 MB at this writing). Remember, however, that you can save documents, images, and other files in a public folder on Google Drive, then link them directly in your site. You can also embed videos from YouTube. Another option is to create additional sites, which can be interlinked. By using all of the available resources, you can stretch your site's memory significantly.
- *Know Your Audience*—it is imperative that you check in with your target market or audience continually. Only your audience can tell you what works and what doesn't. Understand their needs and make corrections to your site in response to their comments and feedback.

The techniques, tips, and advice above should help you and your patrons to create a free, useful web presence with Google Sites. This brief chapter is, however, not an end itself—it is only the beginning of the journey. There are countless features and options available through Google Sites, and only a fraction of them are discussed here. Organizations may find, moreover, that they will eventually outgrow Google Sites. There are a number of other options for hosting and managing websites. Some are free, and even those that charge tend to keep fees quite reasonable (usually $4 to $10 per month). Particularly for organizations seeking to do business online, these services offer functions and security that may well be worth the modest cost. Finally, keep in mind that the pace of change and innovation in website hosting and design is dizzying. It is important to review content, practices, and hosting options regularly to stay current. There is no question that developing even a small website requires a considerable outlay of time and effort. The return to the organization in terms of exposure, communication, accessibility, and impact can, however, be well worth the effort.

Chapter Twenty-Five

Google Books

Shamed by Snobs, a Resource for the Rest of Us

Susan Whitmer

On my first day at the university library reference desk, four students requested *The Divine Nine: The History of African American Fraternities and Sororities* by Lawrence C. Ross Jr. *The Divine Nine* is suggested reading for pledges to African American fraternities and sororities at the University of North Texas (UNT). The UNT Libraries' three copies were checked out and each copy had holds on their catalog records. During my first week at the reference desk, requests for *The Divine Nine* were at the top of the frequently asked questions list.

My reference desk experience happened when I was a graduate student in library information science school where we were taught that Google is evil and Google Books is beneath us because it is easy, only used by the lazy. Google evil? Google cofounders Larry Page and Sergey Brin, created the "Don't Be Evil" corporate motto in order to advance their mission to simplify access and navigation of a confusing abundance of information on the World Wide Web (Vise and Malseed 2006, 178).

Google Books is more flexible and convenient than the library-approved options to assist patrons looking for titles that the library owns but does not have available, like *The Divine Nine.* The library options include interlibrary loan, search WorldCat for local library holdings, and contact the local Greek chapter alumni to request a copy.

Interlibrary loan could take up to two weeks to deliver the title, WorldCat could provide library locations within driving distance, but some students do not have transportation, and contacting local Greek chapters was not appealing to students. Google Books provided the solution.

Using Google Books (books.google.com), patrons can view millions of titles for free. The entire book may not be available on Google Books but an undergraduate student will be able to locate enough information needed to complete an assignment.

Google Books provides various levels of access depending on the copyright agreement Google has with the copyright holder. Books that are in the public domain are books that are no longer in copyright and the entire text is online; this is the Full View option. Limited Preview and Snippet View offer portions of the title. Even the No Preview option provides book reviews, copyright information, locations of the book in local libraries, and purchasing options. Detailed descriptions of access in Google Books follow:

- Full View—the entire book is online. Full view is available when the publisher or author allows the entire book to be accessible because the book is out of copyright or the current copyright holder allows full access. If the book is out of copyright, the patron can download a PDF copy of the book and copy and paste portions of the book.
- Limited Preview—the author or publisher authorized a certain amount of the book to be available online, usually the copyright page, title page, table of contents, and the first chapter with gaps between the pages.
- Snippet View—a few sentences of the book are available online. Books that have snippet view are in copyright, and Google does not have a deal with the rights holder.
- No Preview—basic information about the book: author(s), title, publisher, publication date, reviews, related books. No text inside the book is available in the No Preview mode.

CREATE A GOOGLE ACCOUNT

The benefits of creating a Google account include the ability to personalize Google Books privacy settings, optimize functions, and share books. To create a Google account:

1. In the browser, type: *account.google.com.*
2. Enter required information: name, username, password, current e-mail, birthday, gender, phone number, and location. If you don't have another e-mail account, it is advisable to create one. An alternate e-mail account with another provider is necessary when forgotten passwords need to be reset.
3. Read and check to agree with the Terms of Service and Privacy Policy.
4. Click on Next Step.

5. A confirmation e-mail will be sent to your current e-mail address.

After creating an account in Google, you are ready to customize your Google Books program. You can access Google Books by either going to the apps grid, located in the top-right area, and select the Books icon or type in the browser: *books.google.com*. The customize options include creating public and private bookshelves. The ability to name the bookshelves is a quick reference source when naming them by authors, course numbers, personal interests, and research topics.

HOW TO SEARCH GOOGLE BOOKS

Searching Google Books is easier than searching an online library catalog. Google Books has one search box to enter a title search or a topic search. The library catalog, on the other hand, has multiple search option tabs: Books, Media, Journals, eResources, Subject Guides, etc. The library catalog is cluttered with pull-down menu options: Keywords, Author, Title, Subject, LC call number, Other call number. Most library catalogs are too complicated and return confusing results. Google Books is focused on returning relevant results.

To search Google Books:

1. Put cursor in browser search box.
2. Type: *books.google.com*.
3. In the Google Books search box, type book title or subject information.

Entering *The Divine Nine* in the search box returns a preview copy of *The Divine Nine*. I encourage students who are looking for this title to read the preview copy even though some pages are omitted; it is free and contains the majority of the book's content.

It is important to log into Google Books in order to access the full range of book features available through this program. Simply logging into Google.com does not provide the book's features.

Google Books Search Features

The Google Books search features that are most helpful to scholars include access to book reviews, the ability to locate the book in a local library, and the ability to create a personal online Google Books library using Add to My Library.

1. Text magnification—click on the Magnifying Glass icons to increase and decrease text size.
2. View single or multiple pages—click on the Single-Page icon or Double-Page icon to view documents.
3. Make image full screen—click on the Full-Screen image, four arrows pointing to page corners, to remove menu on the left side of screen. The top toolbar is still visible, only the left toolbar goes away in full-screen view. Click the Full-Screen icon to return to menu view.
4. Create a link—click on the Link icon to paste a link in e-mail or IM, or copy and paste link into a document.
5. Add book to personal library—click on Add to My Library to insert the title into the list of books in your library; your personal library can be public or private. The default category options are: Favorites, Reading Now, To Read, Have Read. To add a bookshelf, click on the red Add a Bookshelf button, enter a name for the new bookshelf, and the new bookshelf is automatically added to the default list.
6. Write a review—click on "Write a Review" to write and publicly display a review in Google Books. All reviews are public, even the reviews of books on the private bookshelves.

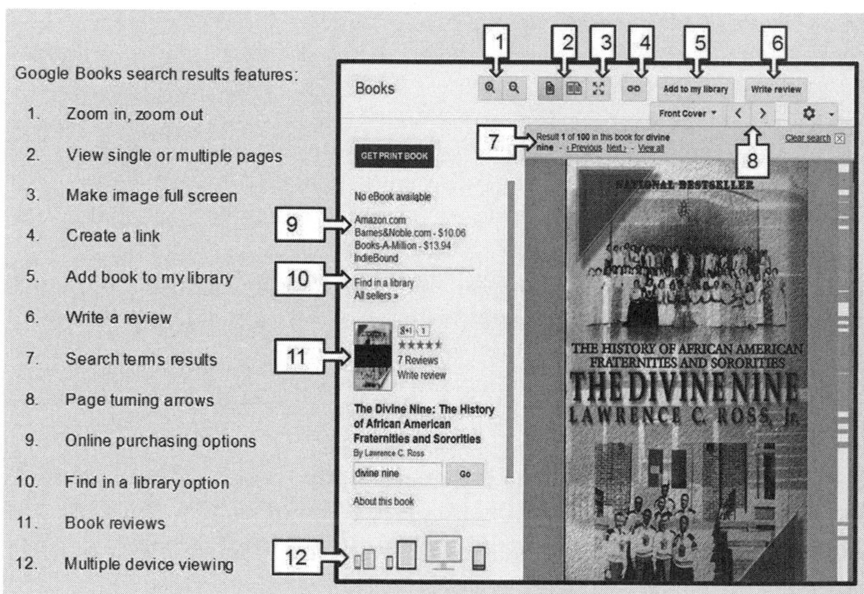

Figure 25.1. Google Books Search Features

7. Number of pages with search terms—click on View All to see how many pages are available to view with selected search terms. The copyright holders determine the number of preview pages available.
8. Page turning arrows—click the right arrow to turn pages forward, the left arrow to turn pages backward.
9. Online purchasing options—click on a purchasing option to order full texts of books. The purchasing options for *The Divine Nine* are Amazon, Barnes & Noble, Books-A-Million, and Indie Bound. Indie Bound provides the option to enter zip codes to locate the closest independent bookstore that stocks the book.
10. Find in a library option—click on Find in a Library to locate libraries that have the title. Google Books searches the WorldCat database, WorldCat is a database that searches the catalogs of libraries worldwide. This option requires the user to enter a local zip code, which returns the holdings of libraries and bibliographic details: abstract, author, format, genre, International Standard Book Number (ISBN), related subjects, reviews, and user lists with this item. The results are listed by location; the closest locations are listed at the top.
11. Book reviews—click on Book Reviews for editorial reviews, User Reviews, and five-star system user ratings. Editorial reviews are criticisms published by professionals in established journals. User reviews are customer reviews.
12. Multiple device viewing—click on the Device icon of choice: desktop computer, laptop, phone, tablet.

Compared to searching a university library catalog, searching Google Books is easy. To search the library's online catalog, the student must find and select the correct tab for books because the current trend in library home pages is to make the article linker tab the default tab. The student then needs to either type "Divine Nine" in quotation marks to ensure that the results return the two words next to each other, the exact phrase, or select "Title" from the Keyword pull-down menu and enter "Divine Nine." These search techniques are not intuitive; the pull-down menus often hide more than they reveal.

Although it is unnecessary to include initial articles when searching specific titles, the only users who know this information are advanced researchers, Library Science students, and librarians. For the novice searcher, the articles *a*, *an*, and *the*, at the beginning of a book title are instinctively included in their search. Google Books is easy to navigate because patrons can use it with natural language, unlike the online catalog, which can feel like you need a master's degree in library science vocabulary to use it.

SEARCH WITHIN A BOOK

To search for words or phrases within a book, start at the search box located in the left column below the book title, enter search terms in the search box. A relevant search in *The Divine Nine* would be searching for Alpha Kappa Alpha, the first Greek sorority established by African American college women. Entering Alpha Kappa Alpha in the search box resulted in eighty-seven snippet results. By clicking on the page number of the snippet, the full page is displayed.

GOOGLE ALERTS

Students looking for *The Divine Nine* may want to do further research into the African American experience in higher education. The students can enter the search term *African American higher education* and receive several thousand titles. However, the students may also want future scholarship on this subject. By creating a Google Alerts for information on African American higher education, students can access titles with perspectives that differ or support *The Divine Nine*.

To set up Google Alerts:

1. Log in to Google.com.
2. Put cursor in browser search box.
3. Type: *google.com/alerts*.
4. Put cursor in Search Query box.
5. Type: *African American higher education*.
6. Select preferences in the Show Options drop-down menu.
7. Click the Create Alerts button.

Logging in to a Google account will automatically fill in e-mail information in Google Alerts. After creating the alert, Google monitors the search terms and will send an e-mail each time a new item is published with the title or subject heading *African American higher education*. The frequency options for receiving alerts are: As-it-happens, Once a day, and Once a week. The preview results are displayed immediately below the query.

GOOGLE BOOKS AND WORLDCAT

Google Books can retrieve most of the books in the world's largest catalog, WorldCat (Chen, 2012). WorldCat is an online catalog of materials from over seventy thousand libraries worldwide. Searching WorldCat requires only Internet access. It is a free, global service, and it is not necessary to have an academic affiliation to use it.

A comparison of Google Books to WorldCat provides similar results when searching for *The Divine Nine*. To locate WorldCat, enter "world-cat.org" in the browser. The WorldCat simplified search form looks like a Google search box with its Everything default option. Entering a book title in the WorldCat search box will return the locations of libraries that own the title; the nearest location is listed first. Clicking on the first result reveals a Google Preview option. Clicking on the Google Preview option reveals the preview in figure 25.1 above.

Google Books and WorldCat share important features: easy searching, book summaries, and library locations. However, WorldCat includes more academic information than Google Books, the kind of information that scholars need for research. WorldCat entries provide detailed catalog data: abstracts, genre, international standard book number (ISBN), linked data, lists of related subjects, and subject headings.

GOOGLE BOOKS AT THE REFERENCE DESK

I frequently use Google Books as a reference source during my daily four-hour reference shift. Wednesday, July 2, 2014, was a typical day at the reference desk where I referred two patrons to Google Books.

The first reference transaction where I used Google Books was with a professor who called frustrated with not being able to access an eBook in our online catalog. After going through the eBooks troubleshooting checklist of trying different browsers, setting up an account with the vendor, and rebooting the computer, it was determined that the book had a broken link. A help ticket was turned in to our technical support staff and the electronic resources librarian. It depends on the problem as to when the link will be resolved. In the meantime, I demonstrated how to locate the title on Google Books, and the professor was happy with the Preview copy until access to the title was resumed.

The second Google Books transaction dealt with an undergraduate who requested Jean-Jacques Rousseau's *The Discourses and Other Early Political Writings*, translated by Victor Gourevitch. The patron did not know how to search the online catalog and confessed that this was her first visit to the library. I wanted to make her first visit to the library a positive experience, so I patiently demonstrated how to find the book in the catalog: start at the Libraries' home page, select the Books & Media tab, select Advanced Search, enter Title: Discourses and Other Early Political Writings, Author: Rousseau.

The one copy of *Discourses* was checked out and the patron did not want to wait for an interlibrary loan, but no local libraries had the title. The patron needed to read this title before class tomorrow. A Google Books search listed the Gourevitch translation as the first result. The patron expressed relief in being able to access this title.

The first transaction with the professor resulted in a discussion about how our Libraries' online catalog broken links make it difficult to do research. In the past, he has either looked for another source or complained to his department chair about the inefficiency of electronic journals and eBooks. This reference transaction was the professor's first exposure to Google Books. He was surprised to see how many titles are available and appreciated my taking the time to explain the navigation tools and search techniques that make Google Books such a useful research tool.

The second transaction with the undergraduate reinforced the need for all students to receive library instruction classes. The librarians have an outreach program designed to teach library skills to all students who attend orientation sessions. The University of North Texas values the research skills taught at the orientation sessions and makes a point to include all incoming groups: freshmen, transfer students, and returning students. However, if a student has no prior library or research experience, starting them with Google Books provides a familiar product as a gateway to navigate the Libraries' online catalog.

COPYING AND DOWNLOADING GOOGLE BOOKS

Only public domain books can be copied and downloaded. Public domain means that the book is not restricted by copyright. *The Divine Nine* is still in copyright, but *The Life and Times of Frederick Douglass* is available for copying and downloading. Portions of text can be copied from Google Books and pasted into a Word document.

How to copy text:

1. Put cursor in browser.
2. Enter books.google.com.
3. In search box, type *The Life and Times of Frederick Douglass.*
4. Select title by clicking once on results list.
5. Locate text to copy.
6. Select scissors icon located above the text.
7. Hold down left clicker, move cursor to select text.
8. Triple-click Selection text, this copies the text.
9. Navigate to document.
10. Hold the Control Key and V to paste text into document.

THE DOWNSIDE OF GOOGLE BOOKS

The downside of Google Books is that Google is a for-profit company that could change their policies in favor of their shareholders over students and scholars. Google Books is free as of this writing in the summer of 2014, but in the future Google's shareholders may want a policy change that is financially beneficial to them.

Another issue is that Google Books provides instant gratification. With immediate access to text, the effort put into searching a well-researched academic library catalog, might be displaced. Trying to focus on reading online text is a challenge with easy access to social media, online games, and other wired diversions. To fully engage with online text, it is necessary to close entertainment tabs or turn off other distracting options.

CONCLUSION

Google cofounders Sergey Brin and Larry Page's corporate motto is "Don't Be Evil." In contrast to my Library Information Science school edict not to search evil Google, this search engine can provide information for scholars that is otherwise unavailable in an academic library. Rare and out-of-print books are available, downloadable, and searchable at point of need on Google Books. "At point of need" means that the student who is working after library hours on an assignment, has the ability to access books that would otherwise be locked up at the library. The finite resources of an academic library can be expanded with the titles available on Google Books.

The major advantage of Google Books is that it is easy to use: book titles are easy to find, the visual aids are easy to navigate, and the site is jargon free. Jargon is a major roadblock for beginning researchers looking for information in an online academic library catalog. Dhawan (2013) argues that undergraduates who are unfamiliar with navigating databases find the process confusing. Vendors design multiple search options to help narrow search results, but these pull-down menus often hide more than they reveal.

Google Books provides an easy-to-use interface with simple navigation tools designed for the user. In comparison, the library user interface is often designed for and by librarians with advanced search skills. Frustrated library patrons need Google Books when the item they require is either unavailable at time-of-need or simply not located at the library. Library staff familiar with Google Books will endear themselves to patrons who are used to Google's user-friendly interface. Library staff who resist familiarity with Google Books will add to the frustration of novice library users and risk losing their support.

REFERENCES

Chen, Xiaotian. 2012. "Google Books and WorldCat: A Comparison of Their Content." *Online Information Review* 36(4): 507–16. doi:http://dx.doi.org/10.1108/14684521211254031.

Dhawan, Amrita. 2013. "Searching Mindfully: Are Libraries Up to the Challenge of Competing with Google Books?" *Library Philosophy & Practice*,1–31.

Ross, Lawrence C. 2000. *The Divine Nine: The History of African American Fraternities and Sororities.* New York: Kensington Books.

Vise, David A., and Mark Malseed. 2006. *The Google Story: Inside the Hottest Business, Media and Technology Success of Our Time.* New York: Random House.

Chapter Twenty-Six

Google Drive for Library Users

Sonnet Ireland

When discussing Google Drive, many users are very unsure just what it is and how it has evolved. Many people confuse Google Docs and Google Drive with each other. Originally, Google Docs was a smaller version of Google Drive (Hamburger 2013). Created in 2006, it did not have all the programs or features that we see today in Google Drive. Over the years, Google Docs grew, eventually offering cloud storage of 1 GB in 2010 ("Upload and Store" 2010). Two years later, in April 2012, Google Drive was released. For quite some time, people used the two terms interchangeably; some still do. However, now Google Docs is actually one small part of Google Drive; it is the word processing option within Google Drive. Quite simply, Google Drive is a cloud storage service that also allows you to create and share documents with free office software.

Google Drive currently offers five major types of software that correlate to other office programs: Google Docs (Microsoft Word), Google Sheets (Microsoft Excel), Google Slides (Microsoft PowerPoint), Google Forms (Microsoft Access), and Google Drawings (Microsoft Paint). Google Drive also offers more advanced software such as Google Apps Script and Google Fusion Tables (experimental phase), but this chapter will focus on the more basic programs mentioned above.

It is important to point out that users do not need to have a Gmail (Google's e-mail) account in order to create a Google account. In fact, when signing up for a Google account, users can choose to use their current e-mail address, regardless of the provider. The benefits for creating a Google account include access to Google Drive, Google Calendar, Google Voice, and Google Bookmarks. It also allows for a more personalized experience when using Google Maps, YouTube, and even Google itself. With that clarified, we'll begin with the feature that makes Google Drive more than just a platform for free software—its ability to serve as cloud storage.

CLOUD STORAGE

Today, technology is constantly evolving at an ever-quickening pace. In the last three decades, computers have gone from using 3.5-inch floppy disks to CDs to Zip Disks to Flash Drives. Each generation of storage offered the user more memory for storage while simultaneously reducing the size of the device ("Timeline of Computer History" 2006). Now, the technological world is enamored with the idea of providing users with endless storage without any device at all—enter the Cloud.

In the 1990s, the average disk offered around 1.44 MB of memory. To put that into perspective, the average song in an mp3 format is 3 MB to 5 MB. The Cloud, however, can offer hundreds of gigabytes—1 GB of digital storage is equivalent to 1,024 MB, roughly the memory of over 700 disks.

Though the Cloud was a vague concept in the 1960s, it was really the bursting of the dot-com bubble that set cloud computing on its revolutionary path (Mohamed 2009). In 2000, Amazon modernized its data centers, making the first breakthrough in the development of cloud computing. It wasn't long before the leaders at Amazon realized the improvement in internal efficiency—and thus the significance of this technology. They quickly set about initiating a new effort to develop and offer this product to external customers. But that was over ten years ago. Now, every user has countless options for cloud computing through a variety of companies—Amazon, Apple, Dropbox, Google, and Microsoft are just a few of the many options out there. Each offers different amounts of storage space for a variety of different (usually reasonable) prices. Google, however, may be one of the best options for library users.

All Google accounts start with 15 GB of storage space. Users can purchase an extra 100 GB of space for as little as $1.99 a month, but the average user doesn't need more storage because only files uploaded to Google Drive take up space. This means that items created through Google Drive do not count against your allotted space. This is only one of many features unique to Google. Google Drive is also an option when users need to save a file but lack the hardware to do so. How many times have users gone to a reference desk in search of a flash drive to borrow or buy? For many systems, this is just not a viable option. Google Drive is one possible solution to that problem. But that's not all that makes Google Drive so useful to library patrons.

FREE WEB-BASED WORD PROCESSING SOFTWARE

The primary attraction of Google Drive to libraries of all types is the lack of software needed. In other words, this service does not require the user to download any software to a computer. This makes it an excellent option for users who do not have access to word processing software at home. Though there are a few open source programs out there for users to download for free, sometimes opening a file from one program in another alters the format of the document. For example, Apache's OpenOffice is a popular and free alternative to Microsoft Office and has many great features. Unfortunately, documents created in OpenOffice can't always be edited when opened with Microsoft Office. Since libraries must often restrict the abilities of patrons to download software, this may mean the user can't do the work they need to at the library. With Google Drive, these issues disappear. Wherever the user has access to the Internet, he or she has access to the document needed.

That doesn't mean the user can't use other software with Google Drive. In fact, users can download any of their Google Docs in a variety of formats, including PDF, Word document, OpenDocument, Rich Text Format, Plain Text, and even Web Page. This holds true for each type of software in Google Drive—multiple downloading-format options means the user has multiple options to further his/her work without hindrance. This means that Google Drive should play well with any type of software the user already has access to.

In fact, downloading a file from Google Drive is pretty self-explanatory. When the file to be downloaded is open in Google Drive, the user can click on File in the menu bar and select Download As. A menu will open to the side with a number of options to choose from. Once a selection is made, the file will download in that format. To download a file from the list of documents, simply check the boxes next to each file to be downloaded. The user can then click on the More button along the top of the screen and select Download. A Convert and Download screen will open with a drop-down menu of options for format.

Thankfully, Google Drive also makes it fairly easy for users to create new documents. As any librarian knows, many users need assistance when using library computers. While we are all happy to help users, it can be frustrating when trying to help patrons navigate software or programs that are not intuitive at all. To create a document in Google Drive, the user simply clicks on the large Create button to the left. This causes a menu to open, leading the user to a list of options. This is where the user can choose what kind of file to create: document, presentation, spreadsheet, etc. It also offers the option of creating folders in which the user's files can be organized.

While Google Drive does not require software to be downloaded, it does offer some downloading options. Users can download Google Drive to their

PCs, Android devices, iPhones, and iPads for free. This download allows users to access and edit their documents without logging in through a browser. Users also have the option to make their files available offline. This allows them to edit their documents even when they do not have access to the Internet.

Downloading Google Drive is easy and takes only a few kilobytes. To download Google Drive, simply go to http://tools.google.com/dlpage/drive and click on the large blue Download Drive button. A drop-down menu will appear with the options: PC, Android Devices, and iPhone and iPad. From there, the Google Drive Terms of Service will appear. Once the user clicks Accept and Install, the program will begin downloading. At that point, it installs like any other program, walking the user through the process as it goes.

Once installed, Google Drive can be used like any other cloud storage software. The user can save files directly to Google Drive. They can also move files into and out of Google Drive with ease. This is one way to add external files to Google Drive. Another option is to upload them through the site. To do this, the user will need to click on the red button located next to the Create button. This button appears with a line and an arrow going upward, symbolizing the act of uploading a document. A menu will appear with the option to upload files or even folders. After a selection is made, a file browser window will open up. The user selects the file he/she wants to upload, much in the same manner that one would use to add an attachment to an e-mail.

SAVE AS YOU GO

An additional benefit of using Google Drive in the library is its auto-save feature. When using any of the software in Google Drive, the software will auto-save your work as you go. Unexpected power outage at the local library? Google Drive ensures that your users won't lose hours of work because of a glitch. It literally requires nothing of the user, which means the user can get into the zone with his/her research without having to worry about remembering to save every so often. This is not a feature offered by any of the other major competitors in cloud computing or word processing software and is yet another feature unique to Google Drive.

SHARING THE WORK AND REAL-TIME COLLABORATION

Google Drive also offers the option to share documents and folders with others. There are three sharing settings available to users: Public on the Web, Anyone with a Link, and Specific People. The "Public on the Web" setting means that anyone can find and access the document/folder selected. "Anyone with a Link" is a little more protected. Only people who have the URL can access the document/folder. Anyone who searches the Internet for it will not be able to locate it. Finally, the "Specific People" setting gives the owner of the document the most control. With this setting, the user provides permission for specific people to find and access the document/folder.

Along with three sharing settings, Google also offers three levels of access: View, Comment, and Edit. With the View setting, those who do not own the document/folder may access it but cannot make any changes in any way whatsoever. Comment, however, allows the users to do just that—make comments on the document without actually changing or deleting the content. This is a great option for the owner who wants input from others without the risk of having their own work lost. The final setting is the true collaboration option; Edit allows those users to make edits to the document just as the owner can.

Sharing a file is actually quite easy. When the user has the document open, he/she will notice a large blue button labeled Share in the top-right corner of the screen. Clicking on that button will open up the settings page. This is where the user can change who has access to the file. There are options to share a link to the file through Gmail, Google+, Facebook, and Twitter. Beneath that is a section labeled "Who Has Access." It is in this section that the user can click on the link "Change . . ." to alter the level of sharing for the file.

Users can also change the settings on their own side with similar setting options: Viewing, Suggesting, and Editing. Viewing allows the user to read or print the document without risk of accidentally changing the content with a misplaced stroke of the keyboard. Suggesting allows the user to insert edits as suggestions. This means that the original content does not disappear and any changes are inserted as suggested edits. This option is popular for collaborative efforts, since it allows the document to change without losing information that may end up staying in the final draft. It also allows the user to show what changes they want made without having to tediously track it themselves. In the past, a considerate member of a group collaboration would highlight or change the font color of words they inserted into a document. They would also have to select text to strikethrough when recommending the removal of words or sentences. This option does all the heavy lifting for those users. Finally, the Editing option allows just that—the user to edit the document without retaining the original information.

To limit his/her own actions on a particular file, a user simply clicks on the icon with the image of a pen, which can be found directly beneath the Share button mentioned previously. When clicked, the pen provides a menu with the three options listed above. If the user wants to find the comments within a file without reading the whole thing, he/she can click on the Comments button immediately to the left of the Share button. This is also where a user can set up notifications and receive alerts every time the document has been commented on or edited.

Just because changes have been made in a document does not mean that the previous versions are forever lost. Google Docs, Google Sheets, and Google Slides all offer a Revision History panel. This means the user can look at what changes have been made throughout the history of the document. To access this history, users simply need to click on File in the menu bar, and then click on "See revision history." Users can also use the keyboard shortcut Ctrl+Alt+Shift+G to access the panel. They can even restore previous versions of the document from this panel with the simple click of their mouse.

If the user is really concerned about losing important content in the file when collaborating with others, he/she can duplicate the file, keeping one intact and private but sharing the other with collaborators. To do this, the user can click on File in the Menu Bar and select Make a Copy. A Copy Document window will open with the option to give the new file a different name. The user can even choose to share the copy with the same people, if he/she chooses. This is also a good option for users who like to create templates for future use. The Earl K. Long Library at the University of New Orleans uses this feature often when working with reference statistics (Ireland and Simmons 2012).

Allowing users to share documents and foster collaboration is becoming a common feature of all cloud computing technologies. Even being able to view previous versions of a document is becoming the norm with these services. The feature that makes Google stand out from the rest is the option to have real-time collaboration on a document. This means that more than one person can work on the document at a time without interfering with one another. One of the nicest aspects of this feature is the ability of each user to see who is currently editing a document. When the document is open, the user can see icons of various colors in the top-right corner; each icon represents the users currently editing/viewing the document. The color of the icon correlates to the color of individual cursors on the screen. This means that each user can see exactly who is editing the document and, specifically, what changes each user is making to the content of the document. This also means that users can turn a Google Doc into a virtual workspace or, possibly, even a virtual meeting place. The possible uses of Google Drive in this context are endless.

RESEARCH

Google is not the first word that comes to a librarian's mind when the topic of research comes up. Actually, Google is rarely considered at all when discussing research, unless it is followed immediately by the word *Scholar*. Even then, many librarians shy away from using Google since so much money is spent on reliable and scholarly resources, whether they be books, journals, or databases. Traditional resources are still vital to researchers, and this chapter does not, in any way, attempt to diminish that. However, there is still room for Google at the table. After all, Google does have its uses when it comes to finding information. If this were not true, then the verb *to Google* would not be so commonplace in our culture. But since the focus of this chapter is Google Drive, that is what this section shall concentrate on.

Google Docs is where most of the tools for research can be found. In fact, under the Tools heading, the user can click on Research to prompt a sidebar to appear. See figure 26.1 for an example. Within the sidebar, there is a search bar where the user can search a topic in Google without having to open a new window or flip back and forth between sites. This search bar can bring up pictures that the user might want to include in his/her research—even filtering the results by usage rights, as shown in the figure. It also has full Google-searching capabilities. This means that, if the user knows how to use Google effectively, he or she can find more resources for the research paper while still writing within Google Docs (Ireland 2014).

The Research function even provides the citation of the resource for the user. Note the citation, circled here in figure 26.2, at the bottom of the sidebar. Clicking on the citation often gives the user the option to find other resources that have cited that particular item. It can even put the citation in the user's document as a footnote or in a bibliography. Granted, there is a more sophisticated option for creating bibliographies in Google Drive.

In March 2014, Google launched Add-ons for Google Docs and Google Sheets (Pinola 2014). EasyBib Bibliography Creator is one of those add-ons. In fact, it is hailed as one of the best add-ons (Klosowski 2014). See figure 26.3. This add-on allows the user to easily search for the bibliographic information of the resources being used—whether a book, a journal, or a website. EasyBib finds the item and bibliographic information needed for a citation; then it actually creates the citation for the user in one of three formats: MLA, APA, or Chicago.

But it doesn't stop there. After the user has added all of the resources being used for the paper, EasyBib will add a bibliography to the document with the simple click of a button. Literally, all the user has to do is click the Add Bibliography to Doc button and all the nightmares over organizing the bibliography are over. Depending on the number of resources, EasyBib may take a few moments to compile the list, but users who have papers to write

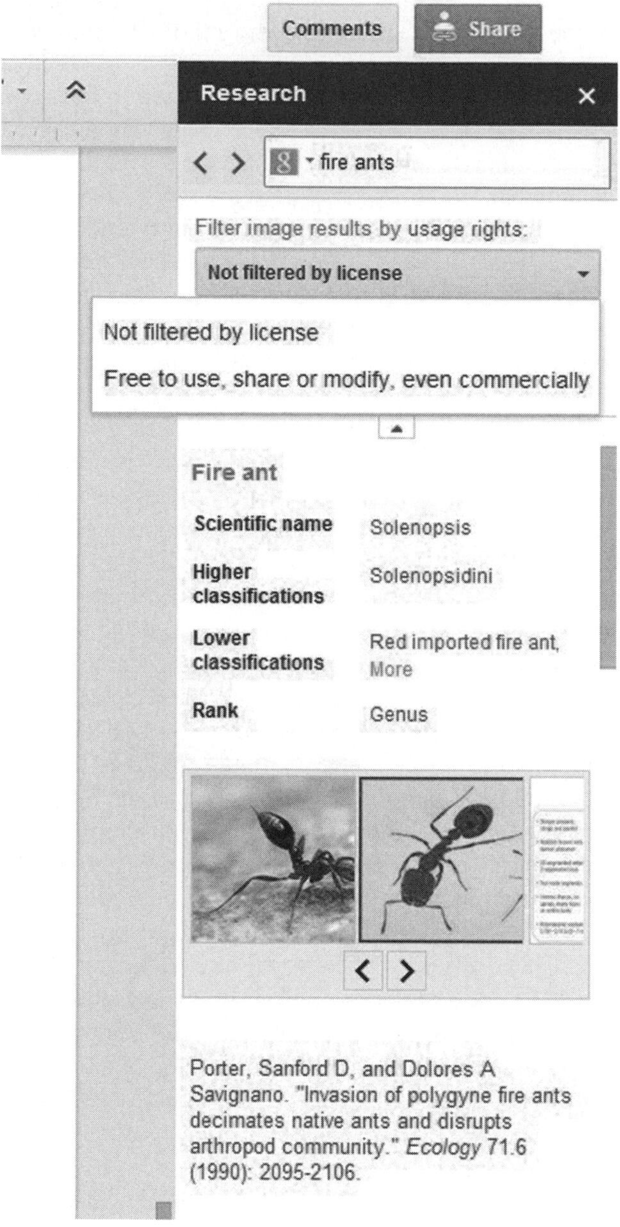

Figure 26.1. Example of Research Sidebar with Filtered Results

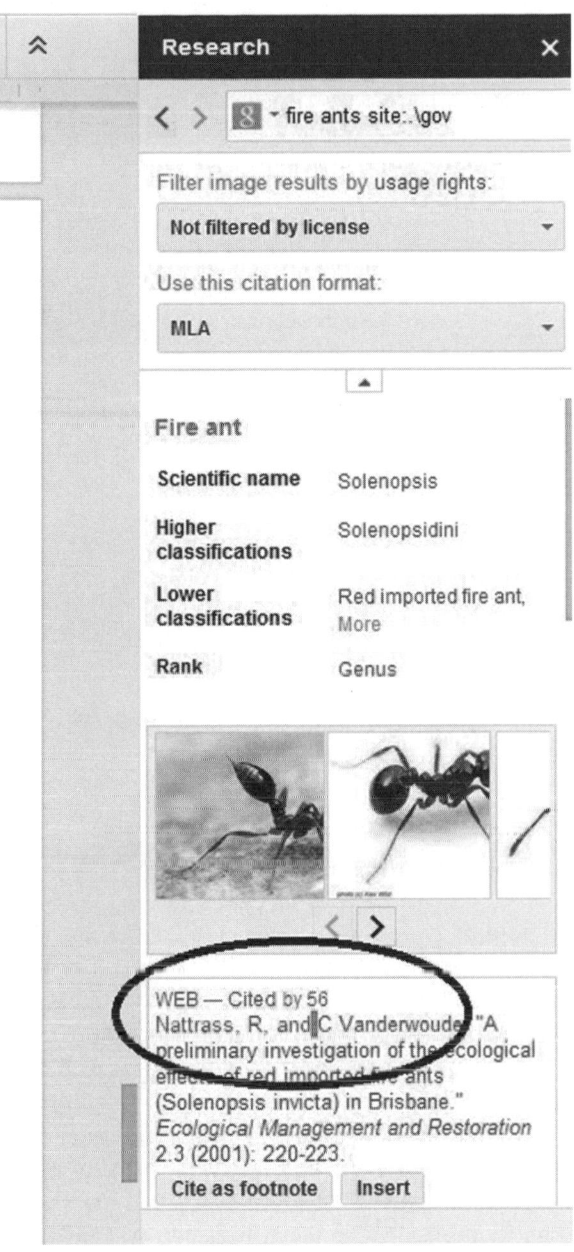

Figure 26.2. Example of Research Sidebar with Citation

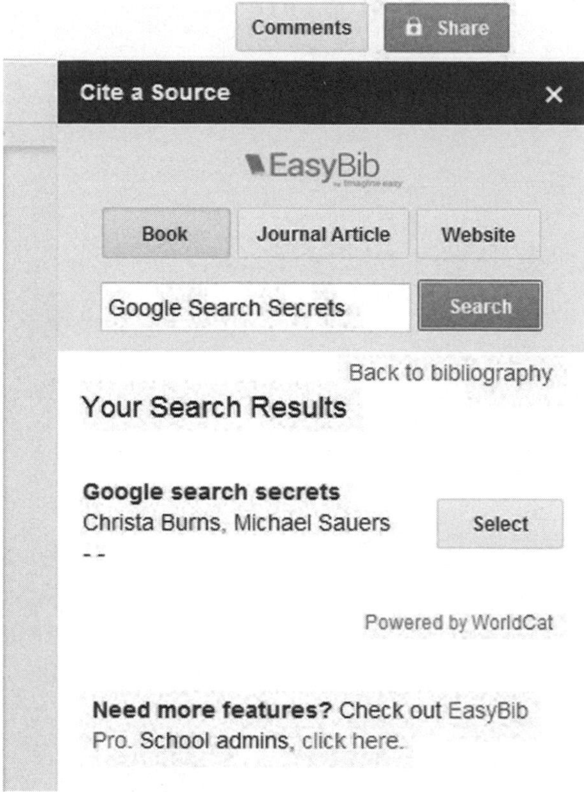

Figure 26.3. Example Sidebar of EasyBib

for school will rejoice over this add-on. Of course, as with any citation management system, there is always a chance for error. It is important to caution users to double-check their citations for accurate format. See figure 26.4 for an example bibliography generated by EasyBib.

To get the add-on (or any add-on in Google Drive), click on Add-Ons in the menu bar. Then click on Get Add-Ons. A new window will appear with images for different add-ons. To find a specific one, such as EasyBib, enter the search term into the search box and hit enter. A list of add-ons connected to that term will appear. In the same row as each add-on is a blue button with a white plus sign (and usually the word Free) on it. Clicking on this button gives the user the add-on. To remove, simply return to the Add-Ons option on the menu bar and click on Manage Add-Ons. A list of add-ons used by the user will appear with the option to manage each one. The user can choose to

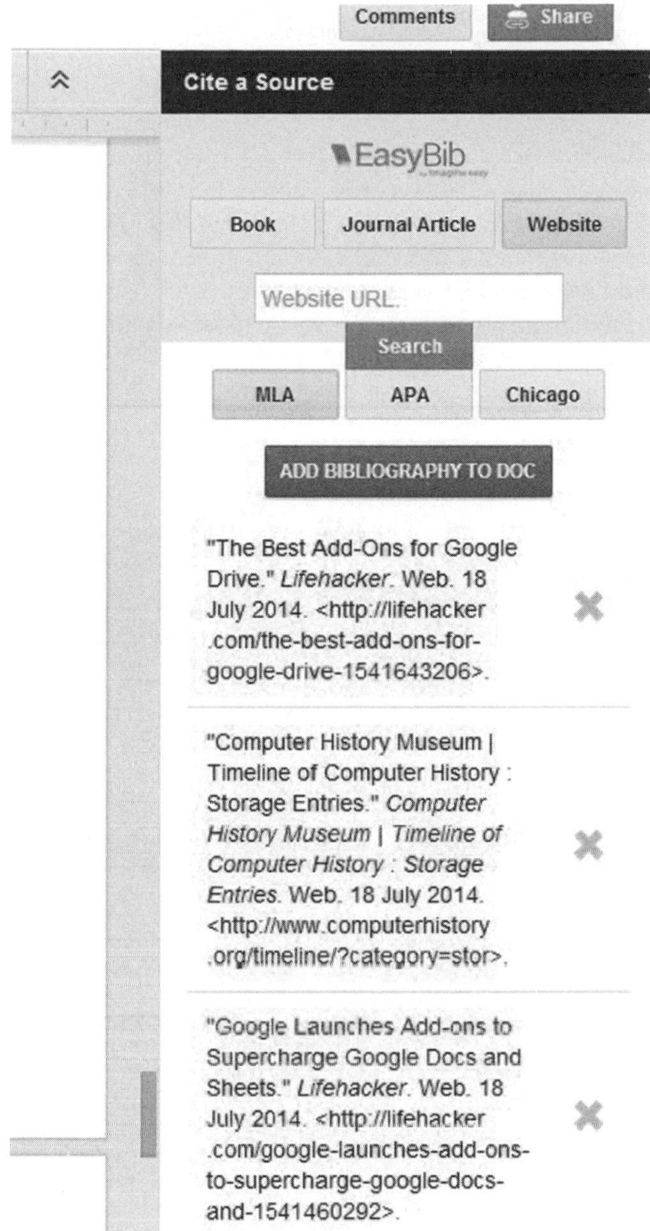

Figure 26.4. Example Bibliography Using EasyBib

only use certain add-ons on a particular file, or he/she can choose to remove the add-on from their account.

CONCLUSION

Google Drive can serve many functions for experienced technophiles and computer novices alike. The fact that it offers free office software with an auto-save feature is amazing in its own right. But Google Drive's ability to act as cloud storage and a free collaboration zone only add to the appeal. While it won't solve all the issues users can have with using library computers, it can minimize some of them. Many librarians will find that while teaching patrons how to use Google Drive they'll fall in love with some of these features themselves.

REFERENCES

Hamburger, Ellis. 2013. "Google Docs Began as a Hacked Together Experiment, Says Creator." *The Verge*, July 3. Accessed July 1, 2013. www.theverge.com/2013/7/3/4484000/sam-schillace-interview-google-docs-creator-box.

Ireland, Sonnet. 2014. "Googling for Answers: The Ins and Outs of Using Google Effectively." Webinar, Learning with LOUIS, Louisiana, April 25. http://connect.lsu.edu/learning_with_louis.

Ireland, Sonnet, and Faith G. Simmons. 2012. "To the Cloud! How to Compile and Analyze Reference Statistics Easily and for Free." PowerPoint, Louisiana Library Association Annual Conference, Shreveport, LA, March 23. www.slideshare.net/SonnetIreland/to-the-cloud-12196936.

Klosowski, Thorin. 2014. "The Best Add-ons for Google Drive." *Lifehacker*, March 12. Accessed July 1, 2014. http://lifehacker.com/the-best-add-ons-for-google-drive-1541643206.

Mohamed, Arif. 2009. "A History of Cloud Computing." *ComputerWeekly*, March. Accessed July 1, 2014.www.computerweekly.com/feature/A-history-of-cloud-computing.

Pinola, Melanie. 2014. "Google Launches Add-ons to Supercharge Google Docs and Sheets." *Lifehacker*, March 11. Accessed July 1, 2014. http://lifehacker.com/google-launches-add-ons-to-supercharge-google-docs-and-1541460292.

"Timeline of Computer History: Storage Entries." 2006. Computer History Museum. Accessed July 1, 2014. www.computerhistory.org/timeline/?category=stor.

"Upload and Store Your Files in the Cloud with Google Docs." 2010. Google Drive Blog, January 10. Accessed July 1, 2014. http://googledrive.blogspot.in/2010/01/upload-and-store-your-files-in-cloud.html.

Chapter Twenty-Seven

Google Finance

Ashley Faulkner

Google Finance is often overshadowed by its rival Yahoo! Finance and other financial websites such as MSN Money. This may be due in part to Google itself failing to promote the site in any meaningful way over the past two years. But however far it has fallen from mass consciousness, this chapter will examine all the many features and the vast store of information that argue for reconsidering this specialized website for both general background research and plentiful instructional opportunities.

Google launched its Google Finance site in March of 2006, with its accompanying promotional blog posting for the first time in June of the same year. In the following eight years, Google Finance has appeared in the news and updated its own promotional materials sporadically, mainly when adding information from new exchanges, or when acquiring real-time (as opposed to delayed) data. When the New York Stock Exchange launched NYSE Real-time Stock Prices in 2008, for example, Google purchased the data for a flat monthly fee and shared in the press coverage of the (at the time) innovative arrangement (Pellechhia 2008). Most recently, Google Finance made the news in June of 2014 when it added Bitcoin prices (and conversion rate to USD) to their coverage (Hong 2014).

We might consider this most recent addition a sign that, while Google hasn't updated its promotional blog or associated Twitter account since 2012, it hasn't forgotten about this offering entirely and users can still find current financial data, if not much in the way of promotional content. Luckily, it's the financial data we're after, and this chapter will explore the breadth of that information and question whether the potential pitfalls of Google Finance might serve as ideal opportunities for library and information and financial literacy instruction.

BACKGROUND

Google Finance is a financial website intended, and best suited for, use as an informational source, rather than for trading purposes. While some of the tools ostensibly lend themselves to use by traders, even these tools are best geared for strictly learning purposes. As long as one bears this tenet in mind, Google Finance provides a robust website that serves users with a solid overview and plentiful background information on markets, sectors, and individual securities. As librarians have long known, Google will likely *not* be the best authoritative or exhaustive source on a given topic (in this case finance) but it *is* a tool even librarians should not overlook when it comes to providing a great breadth of information and plentiful instructional opportunities.

Google Finance breaks its massive data into various categories and tools to help users navigate more efficiently. While the home page may look intimidating, packed with quotes, news, trends, and a visual representation of the market sectors, don't despair! The controlled chaos of this home page provides you with a great opportunity to get an overview of the state of the global market, and you can navigate in a more structured manner from the links in the upper left corner, which outline the structure of the site as follows: Markets, News, Portfolios, Stock Screener, and Google Domestic Trends.

MARKETS

The Market Summary page serves as the home page for the Google Finance site, and it gives users as a solid quick-stop overview of the global market. Top relevant news is highlighted first; if you are viewing the Google Finance page for one of the other nations where the beta product is available (Hong Kong, Canada, the United Kingdom, and China) these will be stories focused on those national markets, but for librarians and users in the United States these will be stories impacting our national market.

As you scroll down the page you'll see quotes for twenty-one of the world's largest markets, currency conversion rates, and a bonds summary on the right. The center of the screen is where you could get a five-second fix on the economy as a whole with the Trends section showing you the most popular company searches on Google, highest price gains/losses, highest market cap gains/losses, and highest volume leaders for the day. My favorite section is at the very bottom: the Sector Summary.

The Sector Summary provides a visual representation of how the entire market is doing, broken into ten component sectors from Energy and Industrials to Financials and Consumer Goods. The percent change for the sectors

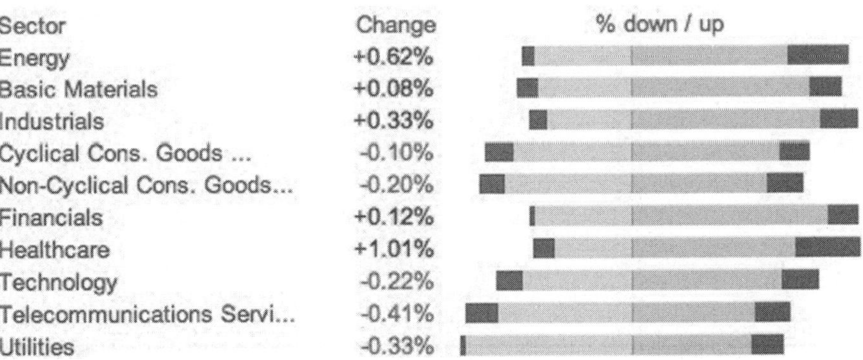

Sector summary

Sector	Change	% down / up
Energy	+0.62%	
Basic Materials	+0.08%	
Industrials	+0.33%	
Cyclical Cons. Goods ...	-0.10%	
Non-Cyclical Cons. Goods...	-0.20%	
Financials	+0.12%	
Healthcare	+1.01%	
Technology	-0.22%	
Telecommunications Servi...	-0.41%	
Utilities	-0.33%	

Figure 27.1. Sector Summary Graphic as Seen on the Market Summary Page of Google Finance: http://www.google.com/finance

is listed with a useful graphical tool to their right; hover over the red and green bars and a pop-up text will tell you the dark red represents the percentage of companies in the sector whose prices are down by more than 2 percent today, the light red the percentage of companies down between 0 percent and 2 percent and the light and dark green represent positive price changes of the same magnitude.

In a quirk of navigation, clicking on the hyperlinked name of one of the ten sectors is the only easily discernable way to get to the individual sector pages. The Market Summary page can be a quick stop for a public library patron or a student just looking to get a five-second fix on the markets at any point during the day. The Sector summary pages are where the opportunities for more in-depth understanding begin.

SECTOR

The first place to which any eye will be drawn is the large chart in the top center of the page. This is an opportunity for the librarian to mention to any patron a few tips regarding both information and financial literacy. You need not go into great detail if finance is not an area you're particularly comfortable with yourself, but at least there is the opportunity to pose questions to the patron. They should notice that the Y-axis is showing percentages, the market sector is shown as one line on the chart, with the S&P 500 charted as a benchmark (the second line), and the date the information was generated is clearly visible. Casual financial literacy instruction might involve discussing what the S&P 500 is and why it's being used as a default benchmark; one information literacy question might be whether data generated the day before, or sometimes a couple of days earlier, is germane to predicting future prices? (As it is not, what other purposes can historical pricing information serve?)

Again, this is not a matter of knowing the answer to these financial questions yourself, but rather, as an information professional, being able to point out what sorts of questions the patron should be asking. And, if they don't know the answers, you can point to where they can go to find that information; we've created here an opportunity for a five-minute library instruction session!

Below the chart will be a list of the companies with the highest percentage price change gains and losses and the most active companies. At the bottom of the page will be a list of all the companies on Google Finance that are associated with the sector. All of the company names are hyperlinked. Click on any to access a company page.

COMPANY

The first thing we see is another large chart. If the patron is unaware, you might point out the differences in this chart, compared to the sector summary chart. We're now looking at price (not percent change) over the time period on the X-axis, and the bottom bars are showing us trading volume. Here is another opportunity for as-you-go information literacy instruction, as you point out that hovering over the names of the various metrics displayed at the top will give users a brief definition; there's a link to the disclaimer wherein users can find a full list of the exchanges, mutual funds, and indexes Google Finance covers as well as whether the data is in real time or delayed, and by how many minutes. How might this information impact our use of the data? The chart is, of course, a rich opportunity for further financial literacy instruction as users can specify settings: they might prefer a logarithmic vertical scale to a linear one, and even add technical indicators from a simple moving average to Bollinger Bands.

If the patron wants to know more about technical analysis, what resources does your library offer on the subject? What books might serve as useful primers and, using your catalog, how can they tell if it's on the shelves or if you'll need to order it from another library? Are there any online databases you subscribe to that might be useful? Other websites? As we prefer Google to work as a whole, the Google Finance site can serve as an ideal diving board into more advanced and authoritative information resources.

The company summary page will also provide a brief description of the company, a list of officers and directors, and external links to analyst esti-

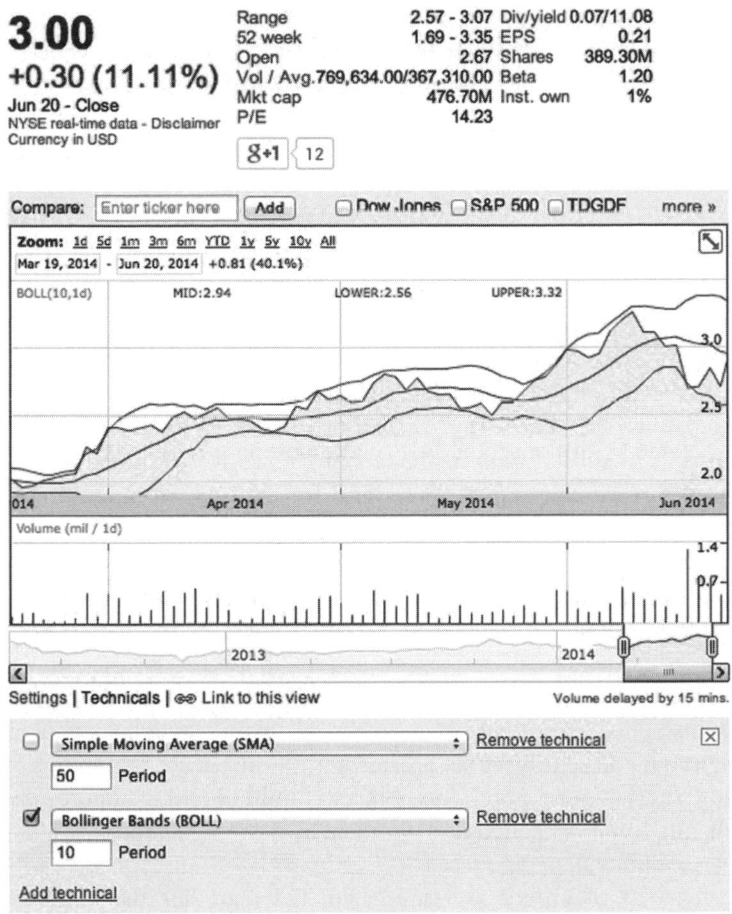

Figure 27.2 Sample Company Summary Page from Google Finance: http://www.google.com/finance

mates, SEC filings online, and so on, but perhaps the most useful internal links are to the Historical Prices and Financials pages. The Historical Prices page will give open, high, low, and closing daily prices going back often to the mid-1990s. If the patron is more familiar with financial analysis, the Financials page provides the Income Statement, Balance Sheet, and Cash Flow Statement with the option of annual or quarterly views.

One information literacy tip you might want to point out here: patrons should always note the currency used in these statements. While more robust financial data sources might have an easy option to convert statements to USD, Google Finance provides these statements in the currency used by the stock exchange on which the currency is traded ("About Portfolios" 2014). If the company you're looking at is traded on the Buenos Aires Stock Exchange, for instance, the financial statements are in Argentine Pesos. This might be an opportunity to introduce your patron or student to the concept of currency conversion rates, if they are unfamiliar, and various sources for currency conversion tables or calculators.

STOCK SCREENER

You can upload a portfolio of stocks directly into Google Finance in OFX or CVS format, but you can also build a portfolio within Google Finance through haphazard searching (there is always the option to "Add to Portfolio" on any individual security page) or using the Stock Screener tool, so I'm going to discuss this tool first ("About Portfolios" 2014). Stock screeners are popular tools in subscription finance databases as well, so learning the basic concept of a screening tool will be useful for patrons that might move on to these more in-depth resources.

A stock screener simply allows the user to derive a list of stocks that meet all of a list of specified criteria. To begin, Google Finance highlights market cap, P/E ratio, dividend yield, and fifty-two-week price change as criteria, but the user can click Add Criteria to access an additional fifty-plus criteria related to price, valuation, dividend, financial ratios, operating metrics, stock metrics, margins, or growth. If a patron is unsure about specific criteria, clicking the name will bring up a brief definition before prompting the user to "Add Criteria" once more. The librarian might take this opportunity again to point out library resources, if more help is needed, such as a financial dictionary or encyclopedia.

When a list of criteria is established, the user has the opportunity to specify a lower and upper boundary for a range of acceptable values. For example, we could specify that we were only interested in stocks for companies with a market capitalization between $10 million and $100 million. We could either type these values into the text boxes or use the slider tool, which gives users a visual hint as to how many companies will fall within a particular range for each criteria. Users can also specify that they're only interested in companies based in a particular nation, traded on a particular exchange, within a particular market sector, or traded in a certain currency. Note that these broad criteria appear above the more particular ranges because specifying these criteria will drastically alter the list of companies you're paring down with the more specific ranges. The list of companies in the screener will automatically update with every additional criteria and range specified.

You cannot add all the companies in your results list to your portfolio at once. This design is likely a reflection of the fact that if you were building a portfolio to trade, it would be unwise to add companies you knew so little about. Google Finance is designed, instead, so that a user must click on an individual company in order to see the option to Add to Portfolio as a button at the top of the specific Company Summary page. Any investor would need at least this summary information before even considering adding this security to their portfolio. For the librarian (who is not using this tool, or encouraging others to use it, for trading purposes!) this is another natural opportunity to discuss with patrons the information needs of traders and in a broader context how to identify an information need for whatever your particular purpose. The Stock Screener results list will give a summary view of all the criteria specified, but the patron should consider what other information he or she would need to make an informed decision about this company.

This is also a potentially visceral way to discuss how much trust a user should truly be willing to put into any one information source. Trusting a single source when you're using it to write a research paper or for a topic of conversation around the dinner table might be one thing, but how much would they trust this source, really, if it was going to be the only information they had to judge a $500 investment? A $5,000 investment? Wouldn't they maybe want to check this information with a second source? How might they go about doing that?

PORTFOLIOS

Whether the user uploads a file or builds a portfolio within Google Finance, the end result is a populated My Portfolio page. You do need a Google account in order to use this feature, but creating an account is fast, easy, and free and many patrons will already have an account thanks to the popularity of Google. Once you have a portfolio composed of a number of companies, this section of the site allows you to track the composite performance of your portfolio and benchmark that performance against the market, and it provides a forum to keep track of your trading transactions, whether these are real or simulated for learning purposes.

My recommendation would be using this tool only for the latter purpose. Like many things on the web, Google Finance is an excellent tool to explore a new topic and learn a basic set of facts and related functions, but it is a poor substitute for more sophisticated information tools when it comes to full mastery of a subject or completing the most complex information interactions. Given the potential for inaccuracies in the financial information available, either through the undisclosed nature of certain calculations (a stock's beta, for instance) or due to potential delays in reflecting major market changes—one financial blogger bemoaned how long it took the site to accurately reflect Twitter's IPO—it would be foolhardy to bank one's financial future on this source (Agrrawal and Waggle 2010; Carney 2013). Of course, this could serve as another opportunity for a librarian to emphasize the preference that an information-seeker never put all his or her trust in one source (see above).

Digressing from the purely informational uses of this section of the site, the ability to chart one's portfolio and benchmark it against the Dow Jones, S&P 500, and/or Nasdaq provides a great opportunity for some financial literacy instruction. Likewise, the Transactions tracker would allow a user to get an idea of what types of transactions can be made in the stock market and the types of information a trader might want to keep track of regarding his or her transactions. For example, you can specify the following four transactions: Buy, Sell, Buy to Cover, and Sell Short. While most users will instinctively understand the first two transaction options, the second two may well provide another learning opportunity. The tracker also allows you to enter in Date, Shares, Price/Amount, Commission, and whether or not the transaction affects the user's cash-balance ("Cash-Linked").

The last major section of the Google Finance site and the last tool we're going to discuss is the Google Domestic Trends pages.

GOOGLE DOMESTIC TRENDS

The Google Domestic Trends tool tracks Google search traffic in the United States under the consideration that tracking real-time search volume may provide unique insight into economic data not typically released until a set delayed date. So, for instance, it might assist us in predicting what actual retail sales volume might be for a given month, before this data was actually released at the end of the given month, if we knew related Google searches had surged early that month (Varian and Choi 2009). Google organizes this information by twenty-seven industry categories and explains in a brief paragraph the search queries tracked and how they composed an index by which to chart volume over time. The chart feature that makes up the dominant portion of these pages allows you to chart these sector-specific search indexes against actual stock performance, the Dow Jones, the S&P 500, and/or Nasdaq. This is one more opportunity for a discussion regarding reading this chart and the nature and limitations of predictive information.

CONCLUSION

Google Finance's disclaimer page includes the following text: "Google does not verify any data and disclaims any obligation to do so" ("Finance Data Listing and Disclaimers" 2014). The paragraph refers specifically to the information on securities provided by the financial exchanges on which they're traded, but indeed it would serve us well to bear this disclaimer in mind, both librarians and all our patrons, whenever we're using a Google tool. Some see the potential limitations to authority and information depth—Google links to outside information sources for all information deeper than a single page into company summaries, for instance—as a weakness of these tools. I see this as an opportunity.

The Google Finance site is an excellent resource for broad market information, securities snapshots, and free tools that might serve as the building blocks for financial analysis and trading skills. Its vast store of stock quotes, historical prices, and financials provide patrons of all financial backgrounds an opportunity to improve their financial analysis skills, while its many interactive charts, its stock screener tool, and its portfolio tool with transaction tracking capabilities allow those who wish to learn more specific financial skills, such as technical analysis or day trading, an opportunity to delve deeper into the mechanics of these fields before they may eventually require subscription-based tools for more authoritative, timely, or flexible information and tools.

Those aspects of Google Finance often derided are precisely what make it such a fertile ground for learning in libraries. With *any* resource, users should always be questioning the validity and timeliness of the information provided. The potential for financial literacy instruction inherent in a financial website is obvious, but Google Finance provides further opportunities for library and general information literacy instruction as users might be prompted to consider the intense timeliness of traders' information needs; the breadth of technical and other chart modifications they can make in this site; where they might search, in the library and beyond, for more information to guide them in these choices; and even the nature of predictive information and determining correlations between broad variables with the Google Domestic Trends tool.

Remember: you don't have to be a financial expert to be an expert at guiding patrons through discerning their information needs (financial or otherwise) and then finding that information. Google Finance is just one free tool in a vast arsenal for library, information, and financial literacy instruction.

REFERENCES

"About Portfolios." 2014. *Google Finance*. Accessed June 15, 2014. https://support.google.com/finance/?hl=en .

Agrrawal, Pankaj, and Doug Waggle. 2010. "The Dispersion of ETV Betas on Financial Websites." *Journal of Investing* 19(1): 13–24.

Carney, Michael. 2013. "Twitter's IPO Reminds Us How Bad Google Finance Really Is." *Pandodaily*, November 7. http://pando.com/2013/11/07/twitters-ipo-reminds-us-how-bad-google-finance-really-is/.

"Finance Data Listing and Disclaimers." 2014. *Google Finance*. Accessed June 23, 2014. www.google.com/googlefinance/disclaimer/#realtime.

Hong, Kaylene. 2014. "Google and Yahoo Finance Now Show the Price of Bitcoin." *Next Web*, June 12. http://thenextweb.com/google/2014/06/12/google-and-yahoo-finance-now-show-the-price-of-bitcoin/.

Pellechhia, Ray. 2008. "New York Stock Exchange Today Launches NYSE Realtime Stock Prices." *NYSE News Releases*, June 24. www.nyse.com/press/1214302996534.html.

Varian, Hal, and Hyunyoung Choi. 2009. "Predicting the Present with Google Trends." *The Latest News from Research at Google* (blog), April 2. http://googleresearch.blogspot.com/2009/04/predicting-present-with-google-trends.html.

Chapter Twenty-Eight

Let's Google "Skepticism"

Easy Searches to Explore Page Rank, Types of Websites, and What Relevance Really Means

Jordan Moore

For my interview at the library I currently work for, the Atlanta University Center Robert W. Woodruff Library, I was asked to prepare a presentation that addressed the importance of information literacy. This was in the summer of 2010, and the BP oil spill in the Gulf of Mexico had been the hot topic for several months. There was much to discuss on this matter in terms of information literacy. This is because the disaster had caused almost as much devastation to the oil giant's reputation as it had the coastline, and amid the accusations of mismanagement and negligence was the revelation that BP was spending money to manipulate Google rankings (funds that could have gone toward recovery, as some news sources pointed out [Mathieu 2010]). It came to light that BP had paid for "sponsored links" to appear at the top of the Google results list when users searched for phrases related to the spill. These links would lead to the "BP Response" page where users could "learn about BP's progress on the Gulf of Mexico response effort" ("Gulf of Mexico Response" 2010). The page contained faux-news stories full of Gulf residents saying how great a job BP was doing with the cleanup, and how there was no ill will between themselves and the company. One story even went so far as to claim that, while the area's seafood business had admittedly suffered, "[m]uch of the region's other businesses—particularly the hotels—have been prospering because so many people have come here from BP and other emergency response teams" (Seslar 2010). What a relief! If one were to get all their information about the oil spill from that first Google hit, one would think that BP had everything under control, and in the meantime, residents were trading in their fishing nets for hotel pillows.

My lesson, of course, did not stop there. I went on to explain that a website's ranking on Google does not necessarily correlate with its value as a resource. Since then, I have found several other searches that show the pitfalls of clicking on that first Google result. I have used these examples to discuss with students the differences between the major types of websites (.com, .net, .org, .edu, and .gov), the purposes they serve, and their reliability. We discuss that if something is found on a .com website, its main purpose is to sell you something, whether it be a product or an idea. The same often goes for .nets and .orgs. I tell students that .edu and .gov sites, while containing reliable and verified information, are also trying to promote their institutions, be they a school, a city, or the entire U.S. government. My students and I also discuss how Google itself is a business; it makes its money by getting users to click on the links it provides, regardless of the quality of the information those links lead to. We discuss the concept that a website's relevance to a search query is just one of the many factors that decide that website's Google ranking.

This chapter provides a variety of searches that demonstrate some of the elements of Google that all users should be aware of, but many are not. They also serve as a means to explain how Google behaves differently from a library search engine. Librarians can expound upon the importance of mindful searching by conducting these searches during library instruction sessions as illustrations of topics relating to information literacy and evaluating resources. This will hopefully allow students to be more responsible Google users, and help them understand why it is often best to conduct research using library resources. A word of caution to the instructor: it is always a good idea to test example searches as close to presentation time as possible, or to make screenshots of successful searches for later use. This is advisable because Google rankings shift constantly due to changes in its algorithm, changes in site popularity, and even changes in users' online habits (more on that later).

BUT FIRST, AN ADWORD FROM OUR SPONSORS

Several of these examples, including the BP example already discussed, mention results that are actually advertisements that companies pay to have appear on results lists. These links are generated by AdWords, Google's online advertising system, and appear at the top, bottom, and right side of Google results pages. In order to get their ads on Google, businesses sign up with AdWords and create a profile for their ads. They set up what search terms they wish to trigger an ad, what geographic locations they wish to cover, and other specifications. Google determines the placement of these ads by two factors: (1) an ad's relevance to the search (its "Quality Score"), and (2) how much the advertiser is willing to pay for the ad space (its "Cost-per-Click" [CPC] bid) ("Google AdWords" 2014). This arrangement creates a tenuous balance between a business's value to searchers and its value to Google. Having a high Quality Score does decrease an advertiser's CPC compared to less-relevant competitors, meaning it is more likely to appear at the top of a results list. Conversely, having a high CPC bid lessens the advertiser's need for a high Quality Score, and could place it above more relevant, but less valuable, advertisements.

Added to this problem is the issue of having these advertisements, especially those that appear at the top of the results list, mixed in with information resources in the first place. Not every Google searcher is looking to buy something, and even if they are, they probably do not want their search results to be shaped by how much businesses are willing to pay Google. So how does one separate these ad links from the links that appear based on their own merit? In the past, advertisements at the top of a results list were placed in a yellow box that denoted them as "sponsored links." See the label pointed to by (a) in figure 28.1 from a 2010 example. In a recent redesign, these links are each designated by a small yellow icon that says "Ad." See the label pointed to by (b) from a 2014 example.

There is currently a debate about whether the name "ad" as opposed to "sponsored link" makes it more obvious that the link is a paid one, or if the lack of visual separation that the old box provided makes it more difficult to distinguish between paid and unpaid links (Miner 2014). What is certain is

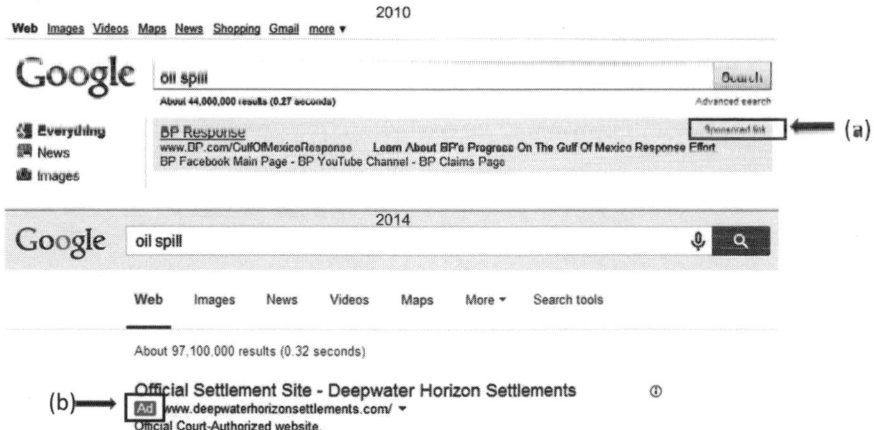

Figure 28.1. Google Indicators for "Ad" Entries

that avoiding potentially disreputable and costly websites not only takes awareness of the existence of paid links, but the ability to recognize them in all their formats, especially when these formats can change at any time. If users are unaware of these sponsored ads, clicking on the first result may be a costly mistake.

THE PASSPORT SEARCH

A few years ago, a friend of mine (a friend who is not at all a librarian living in Atlanta) had her passport stolen during a break-in. She Googled "passport" in a state of panic. So great was this friend's distress that she did not notice the pale yellow box surrounding the first few links, and she clicked on the first one. This led her to U.S. Passport Online, a passport expediting service that allows customers to receive new passports in as few as twenty-four hours. For this service, they charged between $90 and $299 on top of the $170 U.S. State Department fee. It was not until this friend had already paid her expediting fee that she realized she was paying for a service that she did not require. Had my friend looked below U.S. Passport Online, and similar .com sites like Rush My Passport and Passport Visa Express, she would have eventually arrived at the U.S. State Department's U.S. Passport & International Travel site. There, she could acquire her new passport through the normal channels with no additional fee. The experience was a costly and slightly embarrassing mistake, but I share it with my students for two reasons. First, it serves as a reminder to always be aware which websites are paid links. It also illustrates the difference between .com and .gov websites. The .com websites were businesses selling expensive services, while the .gov website offered all the necessary information and services with no upselling. My friend was quite angry at herself for not recognizing her mistake sooner, but I try to cut my friend some slack. She was very upset at the time of her search and was not thinking clearly. And besides, she was not a librarian, who would have known better. (Even if she were, it would simply prove that even the savviest searchers can be steered wrong.)

THE CHANGE-OF-ADDRESS SEARCH

Another person I know (who truly is a different person than the one from the previous example) moved and needed to change his address with the U.S. Postal Service. I saw that he had posted a complaint on Facebook, saying how absurd it was that the USPS was charging him $20 to change his address. Friends wrote to agree that, yes, this was absurd, because USPS only charges $1 for change-of-address services. This person had evidently Googled "change of address" and clicked on the first result. This had caused him to fall for an expensive service that offered the "time saving" deal of filling out and submitting his change of address form for him (never mind the fact that, in order to receive this service, he needed to fill out a form that is nearly identical to the official change-of-address form).

When one Googles *change of address*, there are several of these business websites, all above the legitimate USPS Change of Address page in the Google ranking. The fact that the Postal Service's website (https://moversguide.usps.com) is a .com and not a .gov makes this example trickier than the passport one. To add to the confusion, some of the business sites even feature several trademarked terms such as "US Postal™," "Official Postal™," and "US Postal Address Change™" in their names in order to appear more credible. Users need to look at the fine print of the business websites explaining their inflated pricing and non-affiliation with the USPS in order to see their deception. They then have to look on the USPS "Movers Guide" for the USPS logo and copyright to be sure that this site is the real deal. My friend was annoyed at himself for his mistake, but he was lenient on himself, since the official website was very similar to the other .com that had fooled him. I use this example with my students to discuss how important it is to look at all websites with a critical eye to discern what organization owns that website and what their purpose is. Sometimes the web address, even those critical three letters at the end, are not enough to tell the good resources from the bad.

AND NOW, A WORD ABOUT RELEVANCE

It should be made clear that in the previous examples, many (but not all) of the .com websites listed were marked as advertisements. It is easy enough for searchers to tell themselves that, as long as they avoid these paid links, they are guaranteed to find only the most relevant information, regardless of who is paying whom. These users believe Google to operate in a state of "search neutrality," where the only thing affecting a link's rank is its relevance. But such is not the case. True, Google wants to provide high-quality resources that are relevant to a user's search, but Google's primary interest is providing links that users are likely to click on. As was the case with the ads, the clicks are what support Google's business. That being the case, a link that is more likely to be clicked on by a user will supersede another link that may be just as or more relevant to the user's search.

How Google determines its rankings has always been a topic of speculation and occasional contention. In 2011, Google underwent investigation by the FTC for antitrust activities. Rivals accused Google of unfairly promoting links from organizations with ties to Google and suppressing links from websites associated with competitors. Google's response, as recounted by

Marvin Ammori and Luke Pelican in "Proposed Remedies for Search Bias: 'Search Neutrality' and Other Proposals in the Google Inquiry" was to deny these charges, and to take the opportunity to expound upon the difficulty of true neutrality as it relates to relevance. After all, Google's complex algorithm is making a guess as to what a user wants (a guess with billions of research dollars behind it, but a guess nonetheless). Behind that guess are things like a website's content, but there are other factors, like the website's popularity and the user's likeliness to click on it as extrapolated from his or her browsing history. Ammori and Pelican agree with Google's claim that true search neutrality is impossible because "Search is ultimately an opinion about the likely relevance of some information to a query. . . . If search results had easy 'right answers,' Google would not have to experiment daily and tweak its engine more than 500 times a year" (Ammori and Pelican 2012). What this argument demonstrates, and what I emphasize to students, is that while we may turn to Google for the "right answer" to something, Google is interested in finding what it thinks the "right answer" for us is. The following two examples illustrate this idea.

THE CALENDAR SEARCH

A group of colleagues and I were asked to create a presentation targeting the Atlanta community that would outline good web surfing habits and highlight some helpful community and government resources. We thought that participants would appreciate a calendar of local events. We Googled "Atlanta calendar" and received a very interesting results page. First, Google aggregated a list of events and displayed them at the top of the results page. When we clicked on an event, it became its own Google search, containing information about the event and links to commercial ticket venues such as Ticketmaster and Zvents. There were several sites from other sources, such as the *Atlanta Journal Constitution* and the Atlanta Convention and Visitors Bureau, which offered lists of events and links to purchase tickets, as well. In the midst of their calendars were advertisements for local restaurants and shops. One website, however, was very different from the rest. This was the City of Atlanta's official government page. Under their "Newsroom" tab was a calendar of events that included some of the events listed on the other pages, but also featured municipal events such as town hall meetings and volunteer activities. A large percentage of these events had no cost attached to them, and the ones that did had no associated purchasing link. What some of these listings did offer, however, was information about whether the event would cause street closures or otherwise impact Atlanta's notoriously messy traffic. This search served as another example of difference between for-profit .com sites and the nonprofit .gov sites. The .com sites, Google included, were primarily interested in helping users spend money, at the events as well as businesses around them. The .gov website, however, was interested in the promotion of the city, informing users how they could participate in their local government. The search also served as an example of Google making a judgment call on the website's rankings. While each site did offer a calendar of sorts for the city of Atlanta, Google ordered these resources in a telling manner. It put its own aggregated list of events at the top of the page. It then placed the government's calendar, and the other aggregators (or competitors, as some may call them) underneath that.

THE OBAMA SEARCH

I recently conducted a search experiment to show students how their online habits can have an effect on their Google searches. I cleared a computer's browser history and ran several searches that had liberal connotations. These searches contained phrases such as:

- Democratic Party
- Bill Clinton
- Gun control

I then ran a search for "Barack Obama." The top links were all from the president's official website, the second, third, and fourth of which centered on guns. These links contained phrases like "gun violence," "Newtown," and "gun control." When I was done with that search, I cleared the browser history again and ran searches that had conservative connotations such as:

- GOP
- George W. Bush
- Second Amendment

Again, I searched for Barack Obama, and again, the top links were from the same official website. But this time, the second, third, and fourth results were links to stories about sportsmen associations that all contained the phrase "Second Amendment."

This experiment provided a quick illustration of what is often called the "filter bubble," in which search engines like Google factor in a person's online behavior, including the terms they search for, the websites they visit, and the information they share on social media, in order to determine how likely they are to click on a certain link. In this instance, Google determined that a searcher with liberal political leanings would probably have a negative attitude toward guns and, when researching the president, would want to see things that show he is of the same opinion. The results for that search contained the word "gun violence," as well as the town of Newtown, a place that was recently made forever synonymous with that violence. Google then determined that a person with conservative political leanings would probably see the issue surrounding guns as one of constitutional rights. If they researched the president, they would want resources that promoted the protection of the Second Amendment (note that there was no mention of gun violence in those results). While these links did come from the same source, they speak to two very different sides of the same issue.

I accomplished these results after only a handful of searches. After I showed students this example, I asked them to think about all the Google searches they have done in their lifetime. Then, I asked them to add to that all the things they share about themselves on social media (which Google and many other entities have varying levels of access to). Soon it was easy for them to see how Google could get a good idea of their political leanings, age, economic status, geographic location, and interests. Google takes all this information into consideration in its results list. The immediate outcome is that Google receives more successful clicks from their users. However, this also means that users could potentially receive different information for the same search term. These users are not necessarily receiving the best information on a topic, just the information that Google thinks that they want to see. This has the potential to turn research—something that should expand a person's mind and expose them to new ideas—into something that shrinks a person's view of the world by only showing them what they already think.

CONCLUSION

So, what is the point of conducting all these various searches in the library classroom? In addition to making patrons aware of Google ads, the difference between the purpose and content of commercial and noncommercial websites, the existence of filter bubbles, and the difficulty of neutrality, these searches illustrate the difference between Google searches and library searches. When patrons search Google, they are asking for Google's opinion of what they should see. True, Google wants to satisfy that user's curiosity, but above all, Google wants that user to click on a link (and if that link makes money for Google through AdWords, even better!).

On the other hand, a search conducted through a library search engine brings back results based solely upon relevance. The library does not rank its results based upon how popular those resources are, or which resources the patron is likely to agree with. The library has no "sponsored" resources buying their way onto a results page. When a patron does their research through the library, not only do they have access to a wider array of resources that have been vetted by the academy, they are viewing these resources free from any manipulation on the part of the search engine.

Obviously, there are many occasions when the most sensible thing to do is to "just Google it." Getting directions, finding recipes, figuring out who was that actor in that movie: these are all things that Google makes easy. However, when it comes to finding the best information on a particular topic, users need to be aware of what goes on behind the scenes of a Google search. They need to be aware that different types of websites offer different information and serve different purposes. They must also keep in mind that the website at the top of a Google result list may not be the one that best suits their needs. It could be there because of paid sponsorship, or because its commercial nature provides it with more attention than less glamorous information websites. It may also be at the top because Google has determined that they are more likely to click on it than another, possibly more useful, website. The more patrons know about how Google determines its results, the better equipped they will be to judge which of those results are useful to them. They will also be able to recognize when their research requires results free from bias, either Google's or their own.

A final note: After the fallout caused by their attempt to pass off damage control as news, BP took down their dubious "response page." Although the site lives on in screen captures worldwide, other links have taken its place at the top of Google's rankings. What is the first result for a Google search for "oil spill" in 2014? It is a link to the official page of Deepwater Horizon Settlements, where visitors can submit economic, property, and medical claims.

REFERENCES

Ammori, M., and L. Pelican. 2012. "Proposed Remedies for Search Bias: 'Search Neutrality' and Other Proposals in the Google Inquiry." *Journal of Internet Law*, May 14, 1–48. https://login.ezproxy.auctr.edu:2050/login?url=http://search.proquest.com/docview/1323954886?accountid=8422.

"Google AdWords—Online Advertising by Google." 2014. *Google*. www.google.com/adwords/.

"Gulf of Mexico Response." 2010. BP. www.bp.com/gulfofmexicoresponse.com.

Mathieu, E. 2010. "Oil Giant Mocked for Directing Web Traffic to Site." *Toronto Star*, June 10, sec. B. https://login.ezproxy.auctr.edu:2050/login?url=http://search.proquest.com/docview/366292156?accountid=8422.

Miner, Zach. 2014. "Google's Search Results Redesign 'Makes Ads Less Obvious.'" *Digital Arts*, March 13. www.digitalartsonline.co.uk/news/interactive-design/google-search-results-redesign-makes-ads-less-obvious.

Seslar, Tom. 2010. "In the Heart of America's Coastal Wetlands." *BP Gulf of Mexico Response*, May 24. http://web.archive.org/web/20100703112901/www.bp.com/genericarticle.do?categoryId=9034260&contentId=7062435.

Chapter Twenty-Nine

Navigating Health Information on Google

Why "I'm Feeling Lucky" Isn't Always the Best

Lara Frater

The Pew Research Center, "a nonpartisan fact tank that informs the public about the issues, attitudes, and trends shaping America and the world," estimates that 77 percent of those who looked for health information online started with a search engine, such as Google or Bing, rather than going directly to a health site like WebMD or the Mayo Clinic. The Pew Research Center also estimates that 59 percent of American adults use the Internet to research health issues and women tend to do more searching than men (Fox and Duggan 2013).

Health information on the Internet is often thorough, but may inundate consumers with an immense amount of information. For example, a recent search for *lung cancer* in Google resulted in over 66 million hits. That's a lot of material to sort through, even for a librarian! Yet, it's up to librarians to filter the information and ensure our patrons understand which sites are more reliable. All librarians, regardless of specialization, will have patrons who need medical information, and we must determine the best sources.

This chapter covers the basics of searching for health information using Google. It will help you make sense of domains and determine the good websites from the bad; it will inform you about the reliability of various for-profit websites, and ultimately help you improve health searches by using some important Google resources. Though many of the search strategies discussed can be used with other search engines, all searches conducted for this chapter used Google. It is the most popular search engine in the United States with almost 1.1 billion visits per month compared to Bing, which ranked second, with 285 million ("Top 15 Most Popular Search Engines" 2014). According to a study in the *Journal of Internet Medical Research*, Google "had the best search validity (in terms of whether a website could be opened)" compared to the other search engines (Wang et al. 2012). It seems that Google is the better search engine for health information.

HOW TO SEARCH? WHEN *I'M FEELING LUCKY* ISN'T SO MUCH

In the recent past, Google had a search option: *I'm Feeling Lucky*. When used, only one result was returned, the most frequently used site based on an algorithm called PageRank, created by Larry Page, a cofounder of Google ("PageRank" 2014). Google replaced *I'm Feeling Lucky* with autofill, still relying on PageRank. Now, if you type *American C*, American Cancer Society will pop up before you finish typing, because it was ranked the most frequently accessed site that starts with those words. Other similar or related sites would follow in rank order. Importantly, the site ranked first does not necessarily indicate the best site, but merely the one most viewed (Google even sometimes has ads at the top of the page). So, it is important to apply some simple strategies when using Google's regular or advanced search.

- You must be specific, narrow your search terms, and avoid general terms. For example, if you have a headache and search "Pain" you get the definition of pain but nothing about having a headache. If you want to know what to take for a migraine headache, you will want to search migraine headache treatment, rather than headache.
- Use proper scientific terms rather than lay terms. For example, use gastroenteritis rather than stomach flu or use neoplasm instead of cancer. This is especially pertinent when conducting a search for a consumer who needs more technical information.
- Google offers methods to refine your search. This includes putting exact phrases in quotation marks: *"lower back pain"* in quotes will bring up only those sites with that exact term. A minus sign can be used to remove a word from the search. For example, if you search abdominal pain-WebMD, the resulting sites will include all the sites about abdominal pain but exclude anything from WebMD. If a word is hyphenated, use all variants (e.g., pre-term and preterm) to bring up more results.
- Google spellchecks your search. If you are unsure of the spelling give it your best try. For example: despite spelling *juxtaglomerular* as *juxtglmerlar*, Google was able to find the correct spelling and bring up results.
- In Google's advanced search (http://www.google.com/advanced_search) you don't need to use quotes or minus signs; each is represented by its own search box. You can then narrow down your search by number ranges

(e.g., dates, measurements, etc.), region, file type (e.g., pdf, html, doc), site or domain, and more. Google's advanced search is good for helping to narrow down broad topics.
- Sometimes searching for a symptom rather than a disease yields supplementary information displayed separately, on the right side of the results. There you might find a definition, additional information, links to a site like the National Library of Medicine, and related searches.

USING TOP-LEVEL DOMAINS

The end of a web address is called a top-level domain. This is the .com/.org/.gov that appears at the end of a website address. Sometimes a country code appears after that. For example, the government of the United Kingdom ends with a .gov.uk. For the U.S. Government, most branches end with the .gov but some use .us (Ward 2014). How can top-level domains help when determining the reliability of a health site?

- *.gov* means the domain is run by a government entity (".gov" 2014). Government information in theory provides information free of bias, without an agenda, and presented as facts. It's important to note that government information may have been gathered by nongovernment sources. Moreover, information may not be updated as frequently as commercial sites. Examples of .gov health sites include MedlinePlus, National Guideline Clearinghouse, and FDA.gov.
- *.com* means it is a commercial site (".com" 2014) While anyone can get a .com, it usually means it is a for-profit site and can be funded by ads, other companies (e.g., medical or drug companies), contributions, donations, or product sales. Commercial sites with more sufficient funds could potentially mean more frequent updates. However, .coms may have their own commercial agendas, seeking funding from industries or selling products. Ads on such sites may muddy information provided to consumers. Examples of .com health sites include WebMD, Medscape, and Drugs.com.
- *.edu* means the site is run by an academic institution and is usually funded by it (".edu" 2014). Since these sites are often free of ads, they are more likely free of bias. However, such entities may be supported, in part, by medical industry grants or donations. These centers may post information that will, for example, promote the use of their center. Such information may or may not be medically biased. Examples of .edu health sites include NYU Medical Center (www.med.nyu.edu), Harvard School of Public Health (http://www.hsph.harvard.edu/), and Hardin MD (http://hardinmd.lib.uiowa.edu/).

- *.org* is usually a nonprofit. Technically, the site does not exist to make money (".org" 2014). However, it might have ads to keep up with costs. Even the Mayo Clinic, a recommended Medical Library Association (MLA) health site, accepts ads (Mayo Clinic 2014). Having ads does not, however, automatically mean that the site is not reliable. Some .org sites may be very reliable; others not. Many are run by individuals with their own personal viewpoint on select medical information. For example, an anti-vaccine website might be a .org. Examples of .org health sites include Mayo Clinic, netWellness.org, and familydoctor.org.

WEBMD AND FOR-PROFIT SITES

WebMD is an extremely popular medical information site. It is often the first link listed when you search for medical conditions or diseases in Google, and, as a result, it is frequently chosen to provide consumers with health information. According to Ebiz, a website that tracks Internet usage, WebMD is the most visited health website with eighty million hits per month, followed by National Institutes of Health (NIH), with fifty-five million per month; Medline does not even rank among the top fifteen sites that Ebiz lists ("Top 15 Most Popular Health Websites February 2014" 2014). WebMD may be thought of as a good source for information because of its popularity, but is it? Can a for-profit site be reliable?

For-profit sites tend to have more freedom and therefore more potential to present subjective information designed to support sales and/or profits. As a result, they may not provide the most objective information. In 2011, Virginia Heffernan, in an article in *New York Times Magazine*, criticized WebMD for taking money from drug companies, implying that its health information may be influenced by funding received:

> A February 2010 investigation into WebMD's relationship with drug maker Eli Lilly by Senator Chuck Grassley of Iowa confirmed the suspicions of longtime WebMD users. With the site's (admitted) connections to pharmaceutical and other companies, WebMD has become permeated with pseudomedicine and subtle misinformation. (Heffernan 2011)

In her investigation, Heffernan determined that WebMD supported the use of headache medication more than the Mayo Clinic (Heffernan 2011). WebMD is a for-profit company and, as such, has financing to invest in its users' experience. WebMD does employ physicians to review their content on a regular basis. It's important to note that the MLA's guide *Find and Evaluate Health Information on the Web* (Medical Library Association 2013) currently does not recommend any .com site. Librarians should make sure their patrons know that their doctor, and not a website, should help determine their appropriate health care.

My own mini-experiment comparing WebMD to two nonprofit sites, Mayo Clinic and Medline, and using the search term *gastroesophageal reflux disease* (GERD) came up with findings different from Heffernan. All three gave similar results about medications. Only Mayo Clinic offered a page for alternative remedies (with a large disclaimer).

While .coms may offer some good information, different sites such as FamilyDoctor and the Medical Library Association (MLA) recommend government, hospital, or academic sites as more reliable since they are theoretically free of commercial interests and bias (Medical Library Association 2013). Heffernan recommends Mayo Clinic as a more accurate site to get health information. In the case of .com sites, it might be better to use such sites to collect supplemental health material.

KNOW YOUR PATRONS

I have worked for twelve years in medical and public health libraries. My patrons have included doctors, scientists, nurses, students, researchers, and the general public. If a doctor needs information on a certain subject, be sure to inquire about the extent of detail she is seeking. She may need more than the general information on the disease from the Mayo Clinic. She may need more in-depth research or clinical articles found on Google Scholar and PubMed. A member of the general public may need more general information on their topic, but only your inquiries will help determine what they need. How can you determine what type of information your customer needs? The following are some questions you should ask:

1. If the information is broad, can they narrow the search?
2. What are they looking for? Book? Articles? Studies? Brochures? Handouts?
3. Comprehension level: Are they looking for something simple or complex to read?
4. Is the information for a school project, paper, or just for general reading? A school project or paper might need reliable sources.

Google offers search tools that include a "Reading level" option. To locate, click on Search Tools at the top of the search page. Choose All Results from the drop-down menu, and then Reading Level. It will break up your results based on the reading level of the page: basic, intermediate, or advanced. For example, I searched for "flu." A *basic* search resulted in some general sites with simple language, while *intermediate* included medical articles, and *advanced* brought in even more technical information. You can also sort by date (Search Tools, then All Results, then Date), which provides the most updated information first. You do not want to bring up a site that hasn't been updated in a while as medical information can change frequently.

What else does Google offer to help find reliable health information?

- Google Scholar is a database of scholarly work that includes articles, patents, and case law. Unlike the regular search box, Google Scholar has a drop-down menu in the advanced search. Google scholar is a good supplemental source to gather additional medical articles, especially when you need a broad search. Generally, I recommend using PubMed first to locate health-related articles. You can use the same search strategies as with regular Google. However, with Google Scholar, I recommend always using the scientific term rather than a layman's term (i.e., *gastroesophageal reflux disease* instead of *heartburn*).
- Google Images offers images based on your search term. If a patron for some reason wants to see how a wart looks, Google image is a good place to start. You can search the same way as a regular Google search.
- Google Videos collects videos from the Internet. You can look up talks, lectures, and other videos about health-related information (e.g., I searched *headaches* and got video clips about it).
- Google News offers a searchable site to look up information in newspaper articles. Search as you would in Google, then narrow results by date. You can also create alerts for certain health topics to be delivered to your e-mail.
- Google Books offers some free books, including health-related ones. You can search by title, topic, or browse. Once you find a book, you can search within the contents. You can also save books in "My Library" under your Google account.

TOP-NOTCH SITES FOR GENERAL HEALTH INFORMATION

- Mayo Clinic (http://www.mayoclinic.org/)—This site is my personal favorite and recommended by MLA. Mayo Clinic is a nonprofit medical and research institute whose website offers health information. Mayo Clinic is more informative than other health sites, as it includes multiple pages about diseases and ailments such as definitions, symptoms, causes, risk factors, complications, preparing for your appointment, tests and diagnosis, treatments and drugs, and lifestyle and home remedies. Unlike other sites Mayo Clinic walks the patron through the illness including what will happen at the doctor. A random sampling of ten pages showed that some pages are updated recently while others are updated within one to four years.

- MedlinePlus (http://www.nlm.nih.gov/medlineplus/)—From the National Library of Medicine (NLM), the site offers information on health topics, drugs, and supplements and videos. Each health topic offers a one-page information box. This is especially good for patrons who are looking for basic information and might be overwhelmed by the massive amount of it from the Mayo Clinic. MedlinePlus is also in Spanish (www.nlm.nih.gov/medlineplus/spanish/). According to their website health topics are updated every day, while other parts of the site are updated monthly and annually.
- Family Doctor (http://www.familydoctor.org)—This site includes information on diseases and conditions, prevention and wellness, pregnancy and newborns, kids, teens, and seniors. Family Doctor allows you to search by symptoms, diseases, and age groups (newborns, kids, teens, adults, and seniors). A random sampling of ten pages indicated they were updated within the last four years.
- Healthfinder (http://healthfinder.gov/)—Healthfinder, a U.S. Department of Health and Human Services website, offers health topics, news, and information on health insurance. The main page indicates the site is updated regularly, but their health topics are not dated.
- CDC (http://www.cdc.gov/)—The Centers for Disease Control and Prevention website includes information on diseases, healthy living, and other health issues. A random sampling of ten pages shows they have been updated within the last two years.

The MLA site on finding health information includes a longer list that contains more disease-specific sites. (See www.mlanet.org/resources/userguide.html).

I'M A LIBRARIAN, NOT A DOCTOR: A DISCLAIMER

The term *Cyberchondria* (Fergus 2013) is related to people who suffer health anxiety/hypochondria and use the Internet to obsessively search symptoms. For example, they might search for headache and see that it's a symptom of a stroke and think they are having one, even though a headache is a symptom of many nonserious illnesses. Even I got caught up in self-diagnoses while searching for this chapter.

Health librarians know they will often get calls asking for a diagnosis. Multiple times I had to let patrons know I was not a doctor and could not diagnose them. Resist the temptation to tell the patron they don't have "fatal familial insomnia" (a very rare sleep disorder) because they only slept three hours last night and "Google said so." Offer to send information on the disorder to the patron and gently press them to contact a medical professional, but under no circumstance, even if it's something you know, give a diagnosis. I consider myself an expert at *searching* for medical information but I am not a doctor. Nothing can replace the opinion of a medical professional.

REFERENCES

Allegri, Fran. 2014. "Using Google™ to Find Health Information." *Health Sciences Library*, UNC-Chapel Hill. Modified February 27. http://guides.lib.unc.edu/nchealthinfo-searchenginehealthinfo.

".com." 2014. *Wikipedia*. Modified March 17. http://en.wikipedia.org/wiki/.com.

".edu." 2014. *Wikipedia*. Modified February 16. http://en.wikipedia.org/wiki/.edu.

Fergus, Thomas A. 2013. "Cyberchondria and Intolerance of Uncertainty: Examining When Individuals Experience Health Anxiety in Response to Internet Searches for Medical Information." *Cyberpsychology, Behavior and Social Networking* 16: 735–39.

Fox, Susannah, and Maeve Duggan. 2013. "Health Online 2013." *Pew Research Center's Internet & American Life Project*. www.pewInternet.org/~/media//Files/Reports/PIP_HealthOnline.pdf.

".gov." 2014. *Wikipedia*. Updated March 13. http://en.wikipedia.org/wiki/.gov.

Heffernan, Virginia, 2011. "A Prescription for Fear." *New York Times Magazine*, February 4. Accessed February 11, 2014. www.nytimes.com/2011/02/06/magazine/06FOB-Medium-t.html.

Mayo Clinic Staff. 2014. "About This Site: Advertising and Sponsorship Policy." Accessed April 3, 2014. www.mayoclinic.org/about-this-site.

Medical Library Association. 2013. "A User's Guide to Finding and Evaluating Health Information on the Web." Modified April 2. www.mlanet.org/resources/userguide.html.

".org." 2014. *Wikipedia*. Modified March 3.http://en.wikipedia.org/wiki/.org.

"PageRank." 2014. *Wikipedia*. Modified March 31.http://en.wikipedia.org/wiki/PageRank.

"Top 15 Most Popular Health Websites February 2014." 2014. *eBizMBA*. Accessed February 13, 2014. www.ebizmba.com/articles/health-websites.

"Top 15 Most Popular Search Engines February 2014." 2014. *eBizMBA*. Accessed February 13, 2014. http://www.ebizmba.com/articles/search-engines.

Wang, L., et al. 2012. "Using Internet Search Engines to Obtain Medical Information: A Comparative Study." *Journal of Medical Internet Research* 14(3): e74.

Ward, Susan. 2014. "Top Level Domains." *About.com*. Accessed March 5, 2014. http://sbinfocanada.about.com/od/onlinebusiness/g/topdomain.htm.

Chapter Thirty

Underutilized Google Search Tools

Christine Photinos

Searching with Google can be deceptively simple. Type just about anything into the search box and Google will return results—probably thousands of them. This leads to both frustration and complacency: many users understand themselves to be proficient searchers at the same time that they struggle with large quantities of irrelevant results.

Google makes available a great variety of tools—many of them underutilized—that can improve search precision. Filtering out all irrelevant results is not, in most cases, an attainable or appropriate goal. A more productive goal—and the focus of the strategies described in this chapter—is to increase the concentration of relevant results and promote relevant results closer to the top of the result list.

Below are brief explanations of some especially powerful Google search strategies as well as examples that can be used in instructional settings.

REVIEW SEARCH BASICS

A few search strategies that will strike some as basic rather than as advanced are themselves surprisingly underutilized. Before exploring a variety of advanced search tools that Google makes available, let's quickly review some foundational search strategies.

Search with Phrases

When we enclose search terms in quotation marks, we tell Google that we want results that include only these exact terms in this exact order.

For example, imagine we are searching for a person named Bob Apple. We enter into Google: *Bob Apple.* What do we get? Primarily, we get sources on the game *bobbing for apples*, including the *Wikipedia* entry on *apple bobbing* and the wikiHow page on *how to bob for apples*. To force Google to return results that include only the exact phrase *Bob Apple*, we must enclose the terms in quotation marks: *"Bob Apple"* (note: the presence or absence of capital letters makes no difference here).

Filter Out Unwanted Terms

Often a set of search results will be cluttered with unwanted pages that all have something in common—usually a term or phrase. Pages containing this common element can be filtered out with the minus sign operator. Recipe examples are a quick way to illustrate the power of this kind of filtering. A search for *fries recipe* will turn up mostly recipes for fried potatoes. Now place the minus sign operator in front of the word *potato*: *fries recipe -potato*. This revised search will return recipes for zucchini fries, eggplant fries, beet fries, etc.

Quickly Navigate to Your Search Terms on a Page

Rather than scrolling through long columns of text to find our search terms on a web page, we can use the keystroke combination Ctrl-F (or Cmd-F on a Mac) to go directly to the terms.

For example, if we enter the search term *ACRL digital literacy standards* into Google, one of the top-ranking results will be a page on the ALA website listing the Information Literacy Competency Standards for Higher Education that were adopted by the ACRL in 2000. The title of the web page contains three of our search terms: Information Literacy Competency Standards for Higher Education, Association of College & Research Libraries (*ACRL*).

But what of our fourth search term, *digital*? Why did Google bring us to this page? Rather than scrolling up and down through this document of more than four thousand words hunting for the word *digital* (which appears only once), we can navigate directly to that word by holding down the Ctrl key and tapping the F key (Cmd-F on a Mac). This keystroke combination opens a small search field where we may enter the word or phrase we are searching for on the page.

Let's now turn our attention to some more advanced search strategies, including strategies that are often overlooked or underappreciated by even experienced and nimble Google searchers.

FILTER BY DOMAIN OR URL

Often the type of results being sought will be concentrated on one site or on sites with a particular top-level domain. For example, U.S. government information (government-produced data sets, laws, reports, etc.) will most commonly appear on .gov sites. We can limit a search to .gov sites by using the *site:* operator.

The search query *Texting driving site:gov* will return pages that address the phenomenon of texting while driving from a law enforcement and public safety perspective. Using the *site:* operator with more parts of the URL limits the search further. In the examples below, the first search is the broadest and the last is the narrowest:

texting driving site:gov
texting driving site:ny.gov
texting driving site: dmv.ny.gov
texting site:dmv.ny.gov/tickets

SEARCH USING NUMERIC RANGES

Use two periods (..) between numerals to search for results containing any numbers that fall within a range. For example, suppose we wanted to contextualize an analysis of a present-day political sex scandal by referencing similar scandals from the nineteenth century. The query *"political sex scandal" 1800..1899* will return any page that includes the phrase "political sex scandal" and any number between 1800 and 1899.This approach is more efficient than trying to guess which numbers might appear on relevant pages (e.g., *political sex scandal 1900s*, *political sex scandal 19th century*, *political sex scandal 1801*, *political sex scandal 1802*, etc.).

Numeric range searching can also be useful in finding interpretations of numbers. Examples.

LSAT 155..175
blood potassium 5.1..5.9

Note that there are no spaces before or after the two periods.

We may not know in advance which exact numbers will appear on pages containing the kind of content we are seeking, but range searching allows us to capture results containing our keyword/s and *any* number or numbers falling within a specified range.

SEARCH WITH WILDCARDS

Google wildcard searches are useful for filling in blanks—for example, when we need some help remembering a phrase or phrase structure, or when we have a sense of the type of sentence structure that will include information we are seeking.

Just replace each "blank" with an asterisk.

For example, if we were struggling to remember some common expressions of sympathy, we might search for *"please * condolences"*.

This would lead us to expressions such as *Please accept my condolences* or *Please convey my condolences to* . . . (note: wildcards can be used with or without quotation marks, but this type of search tends to be more powerful and precise when used with quotation marks).

The method can also be used to return other kinds of results. For example, if we were seeking a comparison of online and face-to-face interaction with regard to some activity, incorporating wildcards could potentially lead us to relevant results:

Examples:

*"online * than * face-to-face" classes*
*"online * as * in-person" dating*

The wildcard performs an especially nifty function in Google Maps that is likely to be of interest to anyone traveling or relocating to an unfamiliar area: To get a general sense of a neighborhood, enter an address into the main search field and then enter the wildcard symbol (*) into the Search Nearby field. Google Maps will return an overview of the area including restaurants, parks, government offices, transit lines, community centers, schools, and so on.

FILTER BY TIME RANGE

Just about every Google search environment (Books, Scholar, News, etc.) includes a time range filtering tool. To find this option, run a search, and then click on Search Tools. Change the default Any Time option to a particular time range (e.g., Past Hour, Past Month, Custom Range). This will return results that Google discovered during this time range.

This filter is mostly used to exclude older results, but on some occasions older results may be precisely what is desired.

For example, suppose we wanted to get a sense of the web conversation that was occurring a decade ago around the topic of network neutrality. A search for "network neutrality" within the time range 2003–2005 returns results very different from those for the present. There is a sense in which we are searching the web *as it was* during this earlier period. Of course, content from this period that has since been deleted will not be discoverable in this way. (Here we might turn to the Internet Archive Wayback machine: http://archive.org/web.)

PREDICT—AND FILTER FOR—A SPECIFIC TYPE OF RESULT

The standard advice given to searchers is to try to predict the terms that are likely to appear on the type of page they are seeking, and to search with those terms. Consider, for example, the difference between the search terms *stuffy nose* and *rhinitis*. The denotative meaning of the terms may be the same, but the latter is far more likely to return specialized results. The kind of sources we are seeking should factor into our selection of search terms.

This basic search concept—anticipating what our desired content will look like and crafting the search accordingly—can be extended: we can search with greater precision by making predictions about the *package* in which our desired content is likely to be contained.

For example, if we are seeking data, the kinds of search results we are seeking are likely to be saved in xls (Microsoft Excel spreadsheet file format) or csv (comma separated values file format). We can search for specific file types using the *filetype:* operator.

Examples:

texting driving filetype:xls
texting driving filetype:xlsx
texting driving filetype:csv

For maximum search efficiency and thoroughness, we might add three *filetype:* operators to a single query using the operator OR, in uppercase, to let Google know that we want *any* of these file types: *texting driving filetype:xlsx OR filetype:xls OR filetype:csv*.

Other types of searches can be refined with the *filetype:* operator. For example, suppose we know that the information we are seeking is most likely to appear in the form of a scanned document. We might add *filetype:pdf* to our search query.

Remember that operators can also be used to *exclude* certain types of search results. For example, imagine that a particular search is returning too many results in the form of slide presentations. These results do not represent the amount of detail or depth we are seeking, and clicking through slides is more cumbersome than scrolling through a web page. In such a circumstance, we might add *-filetype:ppt -filetype:pptx* to our search.

Along the same lines, we might reflect on whether our desired content is likely to appear in a short format or a longer format. For example, if we are searching YouTube for a substantial lecture on a topic, there is a good chance the type of video file we are seeking will run longer than twenty minutes. To filter for longer content, we can click on Filters and from the Duration options select Long (>20 minutes).

Consider, for example, the YouTube search query *semiotic analysis*. This search query will pull up primarily short videos—brief introductions, student projects, etc.—while the same query filtered for long-duration content will return primarily lectures and talks by specialists.

SEARCH WITH AN IMAGE

The option of searching with an image rather than with words remains not as widely known as it ought to be. This search method can be used to find other sources in which the image appears as well as similar images. It might be used when a researcher is uncertain of an image's provenance or wishes to learn more about how the image has been rhetorically deployed across multiple contexts.

Both of the aforementioned uses of the tool can be illustrated with political photographs. For example, often a particular image of the president is used in very different ways across the web. Also, while the image is likely to appear without an image credit on many sites, locating the image on a reputable news site will generally reveal a credit (in the case of images of the president, usually a major newswire or daily).

USE GOOGLE SCHOLAR TO TRACK AN ACADEMIC CONVERSATION

As a search environment for academic researchers, Google Scholar has received mixed reviews, but far less contentious is the utility of some of the service's ancillary tools. Particularly noteworthy are three links that accompany search results in Google Scholar: Cited By, Related Articles, and Cite.

Cited By

While end references in a book or article point researchers toward related research from the past, locating subsequent research requires searching for sources that themselves cite the publication in hand. To find sources that have cited a particular source and track scholarly publication on a topic forward in time, use the Cited By links that accompany Google Scholar search results.

This tool is of particular value to academic researchers whose institutions do not provide access to Web of Science (a multidisciplinary fee database that includes a citation mapping tool).

Related Articles

Google Scholar's similarity algorithm will work better with some searches than with others, but the option to explore Related Articles is one searchers should be aware of. The link appears beneath most search results in Google Scholar. Like Cited By, it provides a way to use one source to find others.

Cite

To quickly compile bibliographic notes on sources found while searching Google Scholar, use the Cite links. Available output formats are MPA, APA, and Chicago. Citations can also be imported into citation-management programs such as EndNote and RefWorks. The standard cautions about citation generators apply—they shouldn't be trusted blindly—but this one works well and has the advantage of being available in a massive interdisciplinary index accessible through the open web.

USE ENVIRONMENT-SPECIFIC SEARCH TOOLS

Many search tools—such as custom time-range searching—are available across Google search environments, while others are tailored to specific services. A scan of the Advanced Search options (usually to be found by clicking on the cogwheel in the top right of the screen and then clicking on Advanced Search or Settings) will reveal potentially many helpful tools—even more when we are logged in with a Google account. A few especially noteworthy tools below:

Google Books

Found at http://books.google.com/advanced_book_search, some search options available in this environment include searching by title, author, publisher, and ISBN. Results can also be limited to books for which previews are available.

Entering data into these search fields from the Advanced Search screen will reveal the inline equivalents for each of the operators. For example, if we enter the publisher name *Dover* into the Publisher field on the Advanced Search page, the resulting search summary will look like this: *inpublisher:Dover*.

Knowing these operators is helpful when we wish to combine them with the minus sign operator. Suppose, for example, we were researching the pulp writer Leigh Brackett, whose literary output was enormous. We would have to slog through hundreds of titles (*Queen of the Martian Catacombs*, *Outpost on Io*, *Terror out of Space*, etc.) to pick out the few books containing content *about* her rather than *by* her. The *inauthor:* operator combined with the minus sign operator provides a solution to this problem: *"leigh brackett" -inauthor:brackett*. This search query returns primarily texts about Brackett and her work rather than texts authored by her—for example, *Fifty Key Figures in Science Fiction* and *Leigh Brackett: American Science Fiction Writer*.

The preceding example may be of use to only a small percentage of searchers, but the larger point is to have sufficient overview awareness of environment-specific advanced search tools that one is able to implement them on an as-needed basis.

Google News

Found at http://news.google.com/news/advanced_news_search, options here include searching headlines, specific sources (e.g., CNN), and locations (e.g., a search filtered for the location *San Diego* will return San Diego news sources).

Depending on the type of search we are doing, we may want to exclude blogs and/or press releases from our results. This can be accomplished from the Settings screen.

Google Images

Found at www.google.com/advanced_image_search, in this search environment we can use color and size filters. Images can also be filtered according to usage rights—an option that may be of particular value to those working on public-facing web projects.

Google Scholar

At present, Advanced Search options are accessed by clicking on the down arrow to the right of the search field on the main Google Scholar page (http://scholar.google.com/).

While offering less search functionality than proprietary academic databases, Google Scholar offers a few advanced options that are worth knowing about, among them: searching by author's name, searching by journal title, and searching by article title.

LET GOOGLE CONTINUE SEARCHING FOR YOU

Google Alerts (www.google.com/alerts) provides a way to set up automated searching on topics that are important to us. Once an alert is created, Google will run the search at whatever interval we have designated (Once a Week, Once a Day, or As-It-Happens) and notify us whenever new content is found. This notification can be delivered via e-mail or RSS.

The possible uses for this tool are limitless. Perhaps the most common is reputation monitoring: we can set up an alert that tracks new mentions of our name or organization online.

But any web search of ongoing relevance to us can be set up through Google Alerts to run automatically at regular intervals. Whenever new content is located, Google will let us know.

The alert can be set up to search for specific categories of content—news, blogs, video, discussions, books—or as a general web search.

Of particular interest to academic users will be Google Scholar alerts. Scholar is not a drop-down choice on the main Google Alerts page, but at this writing alerts can still be created from within the Google Scholar service. Run a search; then click on the Create Alert link. Set up the alert to be sent to e-mail or to be monitored as an RSS feed.

RSS is the better method for those who wish to monitor multiple feeds without overburdening their e-mail inboxes. Although Google has—to the great dismay of many—discontinued its own feed reader, Google Reader, there are plenty of other feed-reader services available. Former Google Reader users may want to check out Old Reader (www.theoldreader.com) and InoReader (www.inoreader.com), both of which have interfaces very similar to that of the discontinued Google service.

KEEP CURRENT WITH EVOLVING SEARCH TOOLS

Google search is always changing. Treasured search tools disappear, sometimes permanently and sometimes restored months later. New tools emerge, sometimes with lots of splashy publicity and sometimes with little fanfare. Users for whom web searching is very important—presumably most users—would do well to stay informed about these changes.

Some good sources for monitoring changes to Google search are the following blogs:

Inside Search (http://insidesearch.blogspot.com): Official Google search blog.

SearchReSearch (http://searchresearch1.blogspot.com): Blog of Google search research scientist Daniel M. Russell.

Phil Bradley's weblog (http://philbradley.typepad.com/): Blog on internet searching directed primarily to librarians.

Mashable—Google Search (http://mashable.com/category/google-search): Page on the popular technology blog *Mashable* that brings together news items related to Google search.

Index

About the Editor and Contributors

Carol Smallwood received an MLS from Western Michigan University and an MA in history from Eastern Michigan University. She is the editor of *Librarians as Community Partners: An Outreach Handbook* and *Bringing the Arts into the Library*, both recent ALA publications. Other anthologies for which she served as editor include: *Women on Poetry: Writing, Revising, Publishing and Teaching* (2012); *Marketing Your Library* (2012); and *Library Services for Multicultural Patrons: Strategies to Encourage Library Use* (Scarecrow Press, 2013). Her experience includes school, public, academic, and special libraries, including administration and consulting; she is a multiple Pushcart nominee.

Michael Lesk, professor of library and information science at Rutgers University, New Brunswick, New Jersey, previously worked at Bell Laboratories and Bellcore as director of Computer Science Research. From 1998 to 2002 he headed the Division of Information and Intelligent Systems at the National Science Foundation. He chaired the National Academies Board on Research Data and Information, and his research includes digital libraries, Unix software, and related economic and policy issues. Michael is a Fellow of the Association for Computing Machinery, received the Flame award from the Usenix Association, and was elected to the National Academy of Engineering.

* * *

Giovanna Badia has been working at McGill University's Schulich Library of Science and Engineering in Canada since October 2011. She is the liaison librarian for the departments of Bioengineering, Chemical Engineering,

Earth and Planetary Sciences, and Mining and Materials Engineering. Her responsibilities include answering reference questions, providing instructional services, and collection development. She previously worked as a medical librarian at the Royal Victoria Hospital in Montreal, Quebec, from 2004 to 2011. She also summarizes and critically appraises published articles in library and information science for the journal *Evidence Based Library and Information Practice*.

Cody Behles, an emerging technologies librarian in Memphis, Tennessee, obtained his MLS and MA from Indiana University–Bloomington. His current research is focused on novel uses of technology in academic libraries and scholarly communication. Cody's memberships include the American Library Association, the Library Information Technology Association, and the Tennessee Library Association. His work has appeared in *College and Research Libraries*, the *Journal of Folklore Research*, and the *Middle Eastern Library Association Journal*. Cody serves as founding member and chair for the Midsouth Consortium for Cultural Heritage Organizations and the Library Technologies Roundtable for Tennessee Libraries.

Teresa U. Berry, science librarian at the University of Tennessee, Knoxville Libraries, obtained her MSLS from the University of Tennessee, Knoxville. Previously, she worked at Danville Area Community College Library, Illinois, and the Public Library of Johnston County and Smithfield, North Carolina. A member of the Association of College & Research Libraries and the Reference and User Services Association, she has contributed to *Cooperative Reference and Collection Development* and *Literature Search Strategies for Interdisciplinary Research*. Teresa has been a reviewer for *Library Journal* and was recognized as *Library Journal* Reviewer of the Year for Reference in 2005.

Kayleigh Ayn Bohémier is the science research support librarian for astronomy, geology and geophysics, and physics at Yale University's Center for Science and Social Science Information, New Haven, Connecticut. She received her MLIS in 2012 from the Syracuse University iSchool and participated in the eScience Librarianship Fellow program. Prior publications include a poster abstract on data management plans in the *Joint Conference on Digital Libraries 2011 Proceedings* and an article for *Science and Technology Libraries* on the 2013 Nobel Prize winners in Physics. Her interests include scholarly communication and impact, data management, and embedded librarianship in the physical sciences.

Ashley Krenelka Chase is a library administrator at the Hand Law Library at Stetson University College of Law. Ashley is responsible for the coordination and direction of electronic resources and web page development, as well as coordination of reference and outreach services for the library. Ashley's scholarly interests include the evolution of student and faculty research habits and finding clever ways to incorporate emerging technologies into those habits. Ashley has a BA in English from Bradley University, a JD from the University of Dayton School of Law, and an MLIS from the University of South Florida.

Julie A. DeCesare is assistant professor and head of education and research at Providence College Phillips Memorial Library in Providence, Rhode Island. Prior to this position, she was digital media reference librarian/film studies bibliographer at Boston College 2005–2010. She teaches a blended course, Digital Research Technologies, in the innovative Educational Technology graduate program at Marlboro College in Brattleboro, Vermont. Julie is also a board member-at-large for the Rhode Island Library Association and alumni of Simmons College's GSLIS Program in Boston.

Therese Zoski Dickman, fine arts librarian and an associate professor at Southern Illinois University Edwardsville, obtained a bachelor of music degree (Western Michigan University), AMLS degree (University of Michigan), and MBA. Therese helped create the *KMOX Popular Sheet Music*, the *Louis H. Sullivan Ornaments*, and the *Colket Illustrated Sheet Music* digital collections. She recently chaired the Music Library Association Oral History Committee and has coordinated the MLA, Midwest Chapter's oral history project since 1996, contributing to the "Speaking Our History" series of the *Midwest Note-Book*. Therese has presented about music special collections, oral history, and digitization projects.

Natalie Draper, circulation supervisor at the Ginter Park branch of the Richmond Public Library in Richmond, Virginia, received her MLIS from the University of Kentucky in 2006. She began her library career at age ten shelving books at the Willows Public Library in Willows, California. Since then, she has worked in special collections and archives, and in academic and public libraries. She enjoys developing and teaching digital literacy classes at the branch and exploring ways to incorporate emerging technologies into library programs.

Jennifer Evans is a 2013 graduate of the Masters of Library Science program at the University of North Texas. She is a member of the American Library Association and the Association of College and Research Libraries. She is currently serving as the dedicated project librarian for the grant-funded LEOtrain: Librarians Educating Others Project at Texas A&M University–Commerce. The position allows her to provide on-site technology trainings to the regional rural public libraries of Northeast Texas and their patrons. She also gives conference presentations about engaging library patrons through social media.

Ashley Faulkner is a business librarian and liaison to the Accounting and Finance departments at Texas A&M University's West Campus Library. She graduated with both her MLIS and MBA from Kent State University and her professional goals center around the meeting of these disciplines within the instruction and exploration of information, business, and financial literacy. Her memberships include the American Library Association and the Reference and User Services Association, and she has recently joined the Education Committee of the Business Reference and Services Section.

Deloris Jackson Foxworth is a lecturer in information communication technology in the School of Library and Information Science at the University of Kentucky. She also serves as advisor of the ICT student association. Before joining the faculty at UK, she worked as technology manager at Scott County Public Library in Georgetown, Kentucky. She holds master's degrees in library science and communication. She currently lives in Cynthiana, Kentucky, with her husband, Chris, and their two daughters. Her hobbies include cake decorating and volunteering with 4-H.

Lara Frater currently works as a special assistant in the Office of Vital Statistics at the New York City Department of Health and Mental Hygiene and has worked in libraries for the past sixteen years. She got her MLS from Queens College in 2000. Her chapter "The Lonely Librarian: A Guide to Solo Weeding" appeared in *How to Thrive as a Solo Librarian* (Scarecrow Press, 2012). She is also a writer of fiction and poetry. She lives in Rego Park, New York, with her husband, Jonathan. Visit her at www.larafrater.net.

Rebecca Freeman is an assistant librarian at the University of South Carolina at Lancaster. She graduated with her MLIS from the University of North Carolina at Greensboro, in 2010. Rebecca is mainly in charge of Government Documents, Circulation, and Technical Services, but as with all small institutions she helps a little with everything. When she is not at the library she can be found volunteering at Johnson & Wales University Charlotte Library in Charlotte, North Carolina, baking, or curled up with a good book.

Michael Goates is a life sciences librarian at Brigham Young University in Provo, Utah. He has completed master's degrees in library science (University of North Texas) and integrative biology (Brigham Young University). Michael is a member of the Association of College and Research Libraries and the American Association for the Advancement of Science. Prior to becoming a librarian, Michael worked as an environmental health scientist at a local health department and as a wildlife technician for the U.S. Forest Service and Grand Canyon National Park. Michael regularly helps students at all stages of research and writing.

John Gottfried is coordinator of reference services and business librarian at Western Kentucky University in Bowling Green, Kentucky. He holds an MLS from Indiana University and an MBA from the University of Colorado. His publishing history includes research and commentary for the *Journal of Academic Librarianship*, the *Journal of Business and Finance Librarianship*, and *Reference Librarian*. He is the author of a chapter in *Time Organization for Librarians: Beating Budget and Staff Cuts* (Scarecrow Press, 2013) and is the editor of ALA's *BRASS Notes* and a regular reviewer for *Choice Reviews Online*.

Barbara J. Hampton, a reference librarian at Sacred Heart University, teaches information literacy and research skills to undergraduate and graduate students and presents workshops on information science and technology for educators, librarians, and community groups. She is a member of the Connecticut State Library database advisory committee. From 2007 to 2010, she served as library director for Talcott Mountain Academy, Avon, Connecticut, a K–8 school offering an accelerated mathematics and science curriculum. She earned her MLS at Southern Connecticut State University and her JD at the University of Connecticut School of Law.

Alison Hicks is the romance language librarian at the University of Colorado, Boulder, for French, Catalan, Italian, Spanish, Portuguese, and comparative literature. She has an MA in French and Spanish from the University of St. Andrews, Scotland, and an MSIS from the University of Texas, Austin. Her interests include the integration of critical information literacies into foreign-language learning, and her work has been published in the *Journal of Information Literacy*, *Communications in Information Literacy*, *RUSQ*, and *portal*, among other publications.

Susan Huber is an adjunct faculty member at Argosy University in Eagan, Minnesota, and has taught in the Doctorate of Education program since 2005. She served as the director of the Academic Resource Center until her retirement in 2014. Dr. Huber holds a BS in business from the State University of New York at Albany, an MLIS from Dominican University, and an EdD in Educational Leadership from Argosy University. She has been a member of the American Educational Research Association, the American Library Association, the Special Library Association, and the Minnesota Library Association.

Sonnet Ireland is the head of government information, microforms, and analog media at the University of New Orleans Earl K. Long Library. She also manages the library's chat service and social media accounts. She earned her MLIS from Texas Woman's University in 2008 and is active in the Louisiana Library Association, currently serving as the coordinator for the New Members Round Table and as the vice chair of the Academic Section. She has given numerous presentations on Google and social media. Sonnet is a member of the 2010 Class of ALA Emerging Leaders.

Jesse Leraas is a librarian and teaches as an adjunct instructor at Argosy University in Eagan, Minnesota. He received a BA in English and history from the University of Minnesota, Twin Cities campus, an MLIS from Dominican University, and an EdD from Argosy University. His recent works include a chapter in *The Machiavellian Librarian* (2014) and a doctoral thesis titled "A Phenomenological Approach to Determine How Library Leaders Engage Creativity in Their Decision Making Process: An Integrative Synthesis" (2012).

Melanie Maksin is the Librarian for Political Science, International Affairs, Public Policy, and Government Information at Yale University, where she also leads the Reference, Instruction, and Outreach Committee. She holds an MLIS from the University of Pittsburgh and in 2011 she participated in ACRL's Information Literacy Immersion Program (Teacher Track). She has presented on and written about the role of instruction in academic reference services, as well as instruction and outreach strategies related to government information and other primary sources.

Janna Mattson is the social sciences liaison librarian at George Mason University Libraries in Virginia. She is the coauthor of "STEM Library Services for High School Students Enrolled as University Students: STEMing from Scratch" in *How to STEM: Science, Technology, Engineering, and Math Education in Libraries* (Scarecrow Press, 2013). She has also published in *Virginia Libraries* and the *Journal of Librarianship and Information Science* and has been a presenter at the annual conferences of the American Library Association and the Association of Library Communications & Outreach Professionals. Janna holds an MLS from Queens College.

Jennifer Mitchell is Head of Manuscripts Processing at Louisiana State University. Her major areas of research interest include special collections outreach and access, as well as archival literacy instruction. She holds a master's degree in library and information science from Wayne State University and a master of arts in history from Virginia Tech. Jennifer's professional memberships include the Society of American Archivists and the Louisiana Archives and Manuscripts Association. She also serves as the Louisiana liaison for the Publications Committee of the Society of Southwest Archivists.

Jordan Moore is a reference librarian at the Atlanta University Center Robert W. Woodruff Library. She is the subject liaison for art, fashion, drama, dance, and foreign languages. She also serves as her library's instruction coordinator. Jordan is a member of the American Library Association, the Association of College and Research Libraries, and the GLBT Round Table. Jordan has presented at conferences of the American Library Association, the Atlanta Area Bibliographic Instruction Group, and the Library Orientation Exchange. Jordan is originally from Cincinnati, Ohio, and received her MLIS from the University of Pittsburgh.

Gregory M. Nelson is a chemical and life sciences librarian at Brigham Young University in Provo, Utah. He completed a master's degree in library and information science (University of Southern Mississippi) and a doctoral degree in microbiology and molecular genetics (Loma Linda University). Greg is a member of the American Library Association and the Association of College and Research Libraries. Before becoming a librarian, Greg completed postdoctoral fellowships at the Mayo Clinic-Scottsdale and Massachusetts General Hospital studying protein biochemistry and lymphatic metastasis. Greg teaches library research methodology to a variety of students on a regular basis.

Mary Oberlies is the social sciences liaison librarian at George Mason University Libraries in Virginia. She supports the student and faculty research within the Arts Management program and the School of Conflict Analysis and Resolution. She received her MA in library science from University of Missouri–Columbia and an MA in violence, terrorism, and security from Queen's University at Belfast. Mary's research focuses include regional security studies, international library development and sustainability programs, and the use of emerging technologies to reach library users.

Christine Photinos is an associate professor in the Department of Arts and Humanities at National University, San Diego, California, where she oversees the university's general education information literacy requirement and coordinates the department's graduate specialization in rhetoric. Her research interests include rhetoric and composition theory, digital literacies, and American crime fiction. Her work has appeared in *Clues: A Journal of Detection*; the *Journal of Popular Culture*; *Kairos: A Journal of Rhetoric, Tech-*

nology and Pedagogy; and *Glimpse: The Journal of the Society for Phenomenology and Media*.

Roseann Hara Polashek is head of the Children's & Youth Services Department at the Scott County Public Library in Georgetown, Kentucky. She is the Summer Reading Program coordinator and also works as a reference librarian. Before joining the Scott County Public Library staff in 2010, Roseann worked at the Riverside Branch of the Yonkers Public Library in New York for five years in Youth Services, Reference, and teaching computer classes. She currently lives in Lexington, Kentucky, with her husband, Tim, and their two children. She is an avid reader, and can be found on *Goodreads*.

Steven Pryor is the director of Digital Initiatives and Technologies and an assistant professor in the LIS department at Southern Illinois University, Edwardsville (SIUE). He holds a BS in computer science from SIU, Carbondale, and an MLIS from the University of Washington. Steven has a decade of information technology experience in libraries, including implementing Google Apps domains. He utilizes various technologies to enhance information access, including the 3D printing service he recently launched at SIUE. Steven has presented about mobile technology, open source software, and 3D printing.

Laksamee Putnam, research and instruction librarian at Albert S. Cook Library, Towson University, in Towson, Maryland, obtained her MLIS from the University of Illinois, Urbana-Champaign. She is liaison to the Fisher College of Science and Mathematics. She is an active member of ACRL Science and Technology Section and the Maryland Library Association. Her research focus is on use of emerging technologies and social media to increase science education on all levels, to the public, between scientists, and beyond. In 2012 Laksamee was named an ALA Emerging Leader.

Mary Z. Rose, Metadata Librarian and an associate professor at Southern Illinois University Edwardsville, holds an MLIS degree from the University of Illinois Urbana-Champaign. Mary has collaborated to develop multiple digital collections and exhibitions at Lovejoy Library, notably the *Colket Illustrated Sheet Music* collection, the *Eugene B. Redmond Interviews*, and the *Irving Dilliard Letters* collection. She has conducted two video oral history interviews and presented numerous times about digitization projects, metadata creation, and cataloging. She has been published in *Journal of Library Metadata* (2012) and contributed a chapter to the book *Digitization in the Real World* (2010).

John H. Sandy, head of Rodgers Library for Science and Engineering at the University of Alabama, Tuscaloosa, obtained his MLS degree from the University of Missouri and an MS degree from Southern Illinois University, Edwardsville. He regularly participates in annual conferences of the Special Libraries Association. He is author of numerous technical papers published by Taylor & Francis, Elsevier, and others as well as a major monograph on the bibliography of Nebraska geology. John was awarded the Sci-Tech Annual Achievement Award from the Special Libraries Association, Science-Technology Division.

Rachel Scott is an assistant professor and catalog librarian at the University of Memphis. She obtained her MS in library science from the University of Illinois and has a second master's degree in music. Rachel is an active member of the Tennessee Library Association, serving on the Membership and Conference Committees, and the Music Library Association. She has published articles and reviews in *Music Reference Services Quarterly* and *Tennessee Libraries*. Her research focuses on the intersection of music bibliography and information literacy.

Michael Taylor is assistant curator of books and history subject librarian at Louisiana State University. He holds graduate degrees in library science and history from Indiana University, and has received additional training in special collections librarianship at Rare Book School (University of Virginia), California Rare Book School, and Texas A&M University. His memberships include the Rare Books and Manuscripts Section of ACRL. Previous articles have appeared in *Louisiana Libraries* and *Manuscripts*; he has also written on historical topics for various publications. He regularly speaks to classes on the history of books and printing.

Felicia M. Vertrees, online instructional design librarian at California State University, Northridge, California, obtained her MLIS from San Jose State University, in San Jose, California. Felicia's memberships include the American Library Association, the Association of College & Research Libraries, and the California Academic & Research Libraries Association. Felicia has created more than thirty digital objects and taught virtual classes using Google+. She has held subject-specialty positions as a government documents, business, and psychology librarian. In her spare time, Felicia likes to run, hike with her dogs, and garden.

Andrew Walsh is an information literacy librarian at Sinclair Community College in Dayton, Ohio, who frequently brings Google and other open web resources into his library sessions. He has contributed to *Reference and User Services Quarterly* and *Web Analytic Strategies for Libraries: A LITA Guide*

(2013) and has presented at the Reference Research Forum at the American Library Association Annual Conference. Andrew received his master's in library and information science from the University of Illinois, Urbana-Champaign. He also builds websites and writes fiction in his spare time.

Susan Whitmer, reference specialist at the University of North Texas Libraries, Denton, Texas, obtained her MLIS from the University of North Texas. Susan is a member of the Texas Library Association and is the secretary-treasurer for the Reference and Information Services Round Table. She has given presentations on LibGuides at the Library Instruction Round Table Symposium and the combined services desk at the Texas Library Association conference. Susan has twice been awarded the Star Performer Award at the University of North Texas Libraries.

Andrew Wohrley is the engineering and physics librarian at Auburn University Libraries in Auburn, Alabama. He has additional responsibilities for patents, and data repository duties. In the fields of engineering he has created LibGuides and provided library instruction, and he serves as liaison to the faculty. In the fields of physics he has transitioned the collection to more online sources and has worked to identify and satisfy faculty research needs. He earned his BA at Valparaiso University and his MLS at Indiana University. He is always interested in researching new products from Google that support research.